Homemade Holograms

The Complete Guide to Inexpensive, Do-It-Yourself Holography

Homemade Holograms

The Complete Guide to Inexpensive, Do-It-Yourself Holography

John Iovine

TAB Books
Division of McGraw-Hill, Inc.
Blue Ridge Summit, PA 17294-0850

FIRST EDITION
FOURTH PRINTING

TAB Books is a division of McGraw-Hill, Inc.

Library of Congress Cataloging-in-Publication Data

Iovine, John.
Homeade holograms : the complete guide to inexpensive, do-it -yourself holography / by John Iovine.
p. cm.
Includes index.
ISBN 0-8306-7460-8 ISBN 0-8306-3460-6 (pbk.)
1. Holography—Amateur's manuals. I. Title.
TA1542.I68 1990
621.36'75—dc20 90-44201
CIP

Acquisitions Editor: Roland S. Phelps
Director of Production: Katherine G. Brown

Contents

Introduction

Holograms are becoming rather commonplace today. If you own a credit card, chances are it is embossed with a hologram to prevent forgery. Most supermarkets today use a laser scanning system at the checkout stand to ring up bar-coded merchandise. These systems employ a laser with a holographic lens element. Hence, most people either possess a hologram or have seen the technology in use firsthand.

Although holography's roots trace back to 1947, the evolution of this technology has been slow. There are still many holographic frontiers to conquer and many technologies to develop. The "Thomas Edison" of holography hasn't yet made an appearance. This is not to say that holography hasn't advanced to a point where you can't use the existing technology for purely artistic pursuit, because it has. In fact, it has been the holoartist who has kept the interest in holography alive.

During the early 1970s, much of the science research in the field was abandoned because technological problems weren't as easy to solve as many first thought. The economic pressure these difficulties caused forced many companies to retreat or abandon their holographic research. But much has been accomplished since, and many companies are back into the fray. For the innovator, there is still much more room to experiment than has already been explored.

Some technical and instructional books leave individuals with the impression that their material is definitive. While a considerable amount of information has been amassed over the years, that could quickly overwhelm beginners in the field. The basic operating procedures for shooting and developing holograms remain simple and direct. The amassed technological information usually deals with optimizing a particular detail of the holographic setups or equipment rather than providing a new innovation in the field.

This book provides a solid foundation in holography. You can shoot and develop various types of holograms, starting with the most simple single-beam setups and advancing to more elaborate ones. The equipment is innovative. For instance, there is no need for a sand table, which has been the main type of table for amateur and some professional holography for many years. Sand tables can be a major impediment for hobbyists who want to shoot holograms (in fact, it was the main reason that stopped me from pursuing my interest in holography for years) because after all, how many people have the room to support a 500- to 2000-pound monstrosity in their home? It certainly isn't portable, and you can't even put it away after you are through working with it for a day. For apartment dwellers, it's totally out of the question. You would have to go as far as checking the building structure to see if your floor could support that amount of concentrated weight in a small area. Instead, this book contains plans for a small, lightweight table (20 pounds) that is easy to construct and can be set up or broken down in a few minutes for easy storage and retrieval. In addition, there are plans to construct suitable lasers for holography for $100 and simple magnetic optical mounts and component holders designed for use on our lightweight table.

In the end, if you want to pursue advancing further into holography, you will have a foundation to build upon. At the end of this book, some areas of exploration bring you to the threshold of current technology. The possibility for new discovery is eminent. The Sources appendix in this book indicates publications and organizations that can keep you abreast of current information in holography. Many organizations publish books that are compilations of articles by holographers from the current year. Many of these books are a goldmine of tips, techniques, and procedures in holography.

Holography can be a very simple process. You don't have to be an expert in all of the technical areas of holography to shoot and develop first-rate holograms, just as it isn't necessary to learn electronics to watch television or listen to a radio. The technical understanding can of course provide a greater understanding of the processes involved and are of benefit for anyone who wants to pursue developing holography as an art or science.

This book separates the basic holographic processes from the more advanced technologies in an effort to provide you with the option of either skimming the surface of the technology while making holograms or delving as deeply as you want to in the technological sea. If the contents of this book don't provide you with the information you seek, it will direct you to where that information can be found.

Throughout this book, heavy emphasis is given to information contained in the appendices for two reasons: first, the information in the appendices is typically short and discrete, such as the section on eye safety with lasers. Although this information could easily fit into a larger, more general chapter, it would be more difficult to locate. Second, some information doesn't neatly fit into any one chapter and could therefore be referenced as needed such as sources and additional projects.

Chapter **1**

What is holography?

Holography, like photography, is a technique that produces an image on film. The method by which a hologram is recorded and produces an image is very different than conventional photography. Standard photography creates a two-dimensional image on film of a subject. The film image is called a *negative*. The two-dimensional image on the film is of a single unchangeable viewpoint. The third dimension, or depth dimension, of the subject is collapsed onto the plane of the film.

Holography, on the other hand, produces an image on film called a *hologram*. The hologram records all of the visual information of a three-dimensional scene, including depth. Subsequently, this allows you to not only view the original scene in three dimensions but to also view the scene from many angles by moving your head side to side or up and down. Essentially, you can look "around" objects in the hologram.

HOLOGRAPHY VERSUS PHOTOGRAPHY

To gain an appreciation of holograms, begin by noting the differences between holograms and photographs. Understanding the differences should provide a rudimentary understanding and introduction into holography. All subjects touched upon here are covered in greater detail in later chapters.

Film

As mentioned, holograms, like photographs, begin with forming an image onto photographic film. The photographic film required for holography, however, needs a much higher resolution than standard photographic film. Holographic film typically has a resolution of 2,500 to 5,000 lines per millimeter (10^{-3} meter), in contrast to standard photographic film, which has about 200 lines per millimeter. The higher resolution is achieved by using smaller grains of the photosensitive silver in the emulsion. The smaller grains are less sensitive to light and decrease the "speed" of the film substantially. This in turn increases the required energy and/or exposure time. You will better understand the need for holographic film to have such a high resolution after reading the section on recording interference patterns.

Photographs Figure 1-1 shows a drawing of a simple box camera. The image formed by the lens onto the film is a real image, and this real image is recorded onto the film. The resulting photograph is a two-dimensional representation of the subject. The third dimension (depth) collapses onto the film plane. Lighting for photography can come from any common light source such as the sun, electric lights, or flash tubes.

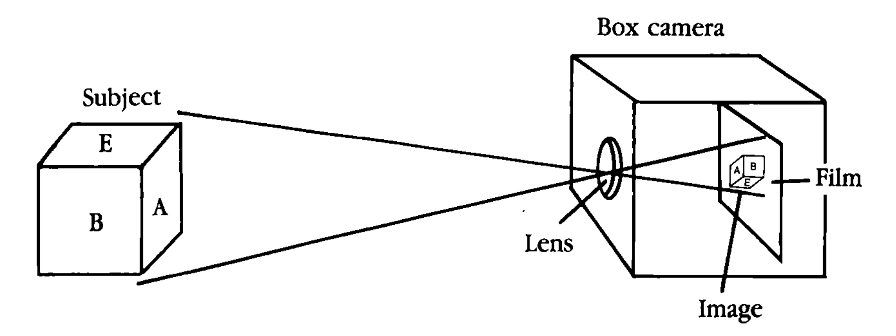

1-1 Simple box camera.

Holograms Holography does not record the subject's image the same way the camera does. Holography essentially records the interference pattern of light generated from a reference beam and reflected light from the subject (object beam). See FIG. 1-2. The light source required for holograms must be *monochromatic* (single light frequency) and *coherent* (wavelengths in phase). Laser light fits the bill admirably.

In the illustration, a typical split-beam holographic setup is used. It is possible to record holograms using just a single beam. In fact, the first holograms presented in this book are single-beam setups, but in an effort to present a diagram that clearly illustrates the interference pattern created and recorded, a split-beam setup is clearer.

When the resulting hologram is properly illuminated, the resulting image is a true, three-dimensional image of the subject. It is important to clarify at this point that the resulting three-dimensional image is not a trick, illusion, or psychological effect. The image is truly three dimensional. This can be proved by photographing the holographic image with a camera set at different angles and perspectives. The resulting photographs will show the subject from different angles as if the subject were actually in front of the camera (see FIG. 1-3).

Figure 1-3 demonstrates the parallax of the hologram. The vertical parallax is from top to bottom, and the horizontal parallax is from left to right. *Parallax* is a term referring to the viewable angles of the subject in the hologram. With a photograph, looking at the subject from an angle just creates a foreshortening of the flat image, so the image remains at the one viewpoint alone.

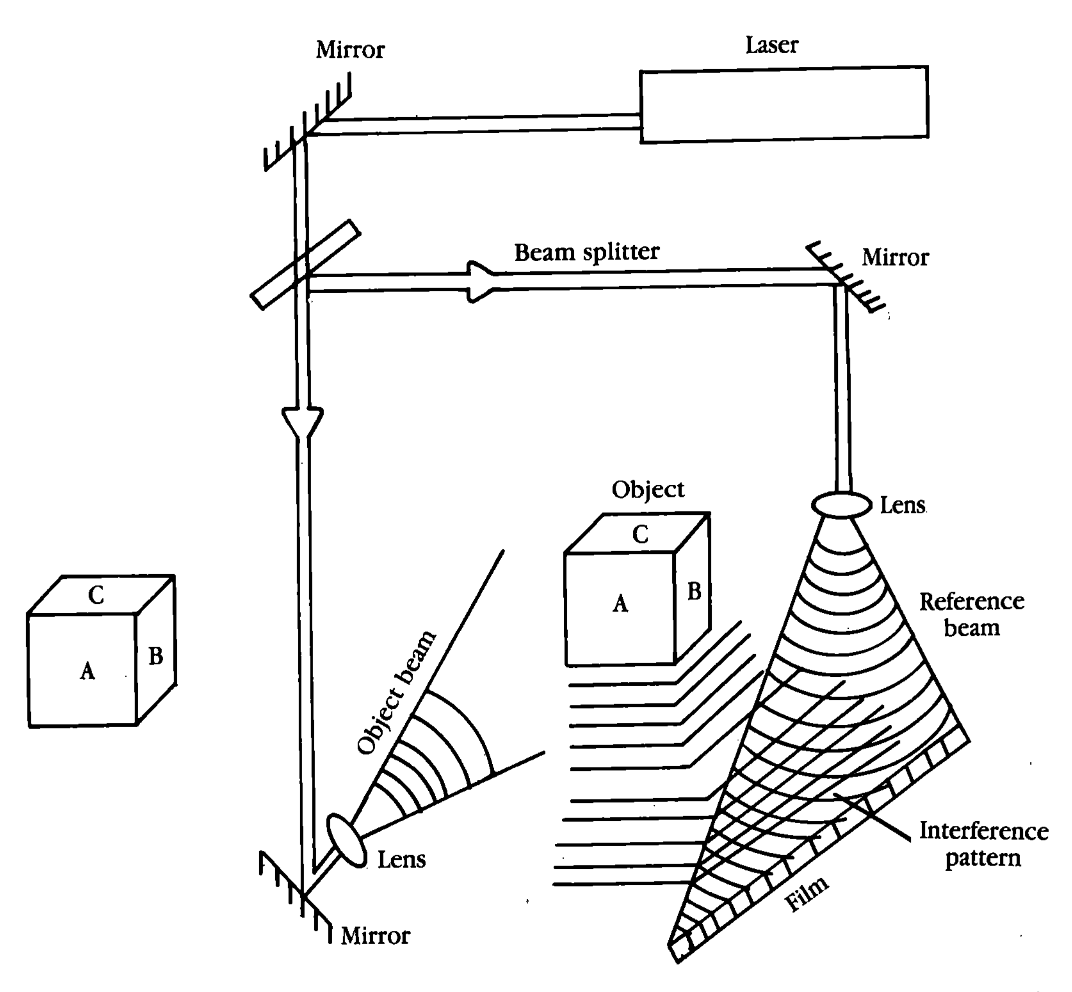

1-2 Overview of recording a transmission hologram. The interference pattern is created by the reflected object beam and the reference beam and is recorded onto a film plate.

Redundancy

If a hologram is broken into small pieces, the entire image would still be viewable through any of the broken pieces. This becomes easier to comprehend if you look back at FIG. 1-2. Imagine that the holographic film becomes a window with memory when it is exposed. When any object from any viewpoint that is visible through the window (holographic plate) is exposed, it can again be viewed when the image is reconstructed. All perspectives of the subject vertically and horizontally have been faithfully recorded in the hologram, so much so that if you covered the hologram with black paper and put a small peephole in it, you could still view the entire subject through the small hole as if you were looking through a peephole in a window. (See FIG. 1-3.) Where you place the peephole on the window (hologram) determines from which perspective you would see the subject.

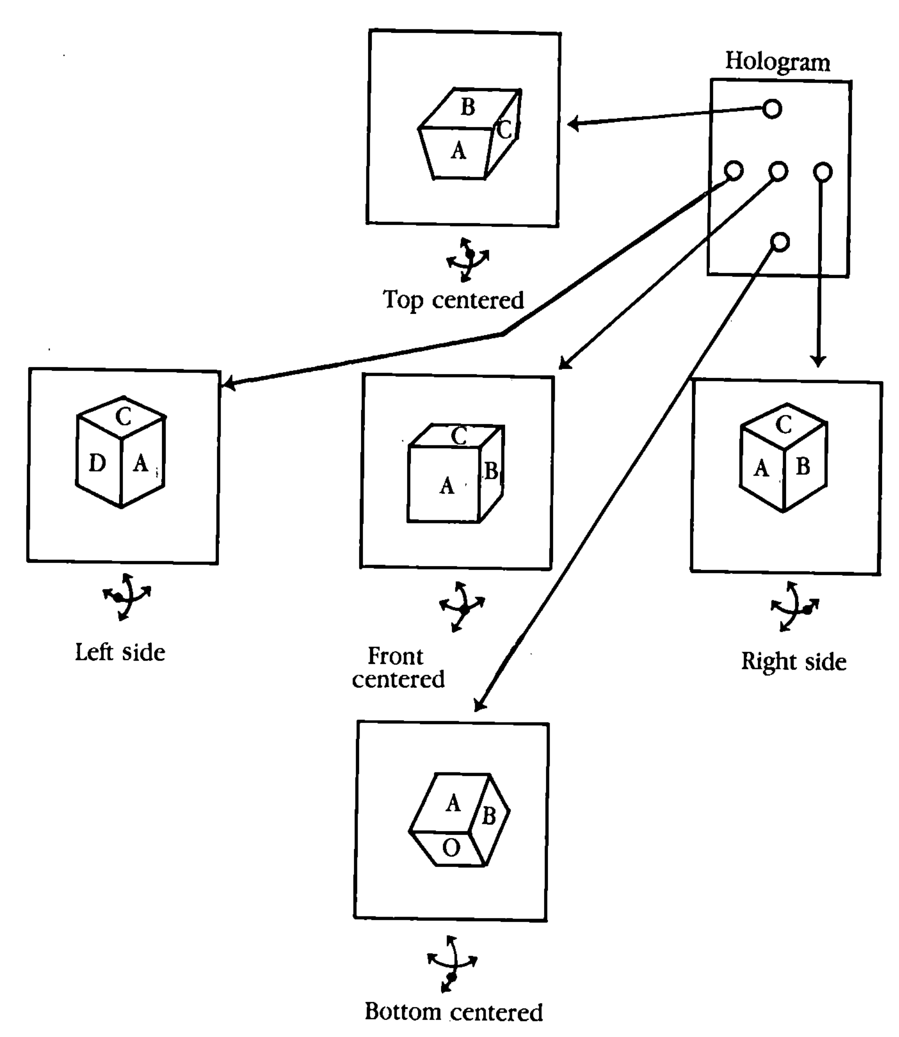

1-3 Parallex views recorded in a hologram.

Copying

The original piece of film produced in a camera is a negative. Photographs (positives) are printed from film negatives. During printing, the photographs from the negative can be enlarged or reduced from the original size of the negative. Also, the negative can be used to easily produce numerous copies of the photographs.

Holograms are the original pieces of film. They are not referred to as negative or positive images as in photography. Since the original piece of film is the hologram, it is not printed to another media for viewing. Therefore, when you want to make a copy of a hologram, there are some limitations when compared to conventional photography. One limitation is that the image can't be easily enlarged or reduced as in conventional photography. Copies are not as easily reproduced. To make a copy of a hologram, the original hologram serves as the master. The three-dimensional real image from the master is projected and holographed onto a second plate (copy).

There is also a method of stamping inexpensive foil holograms. These holograms are the type typically used on book and magazine covers. The initial investment in the plate and the minimum quantity of reproductions required to make a press run keeps this beyond the reach of most amateur holographers. This is a high-speed, high-volume reproduction method.

Chemistry

The chemistry required to develop holograms is not any more complex than standard black-and-white photodeveloping chemicals. In some cases, they are the same. The procedure of developing holograms is also similar to developing black-and-white film. Neither is very complex and should not present any problem to budding holographers.

The bottom line is that good holograms can be produced with a minimum investment of four basic chemicals. These chemicals are mixed with water (preferably distilled water) to make three stock solutions. Two of the stock solutions are mixed into one tray, which is the developer. The other stock solution is the bleach and is placed into its own tray. A tray containing water is placed between these two trays and is used to rinse the developer off the plate before placing it in the bleach. This step preserves the bleach so you can reuse it. The complete procedure is outlined in chapter 6, "Developing Holograms."

MAKING HOLOGRAMS

A working model of the creation of a hologram that can be helpful when actually shooting holograms is the *geometric model,* devised by Dr. T. H. Jeong. The geometric model is easier to understand than the *diffraction model* because it doesn't require mathematics to be understood. It provides a good foundation for the technician as well as the artist. Although the geometric model is limited in some applications, it can carry you a long way into the field of holography.

Light

Visible light, as we commonly know it, is composed of all the colors in the rainbow. We can observe this phenomenon by passing a narrow beam of sunlight through a prism. The prism separates the light into its component colors and projects them.

In the late 1600s, Christian Huygens proposed that light propagates as a series of waves and that any point on the advancing wave is capable of generating or becoming a new source of waves. Figure 1-4 illustrates these points. Light waves can be visualized as a series of waves that are created when a pebble is dropped into a pool of still water. The distance between two adjacent peaks is defined as the wavelength and is denoted with the Greek symbol lambda (λ). As the waves impact (illuminate) the hole in the barrier, waves are generated from the hole in the barrier as if it were the source of the waves. The wavelength of the waves emanating from the hole are the same wavelength of the waves impacting the barrier.

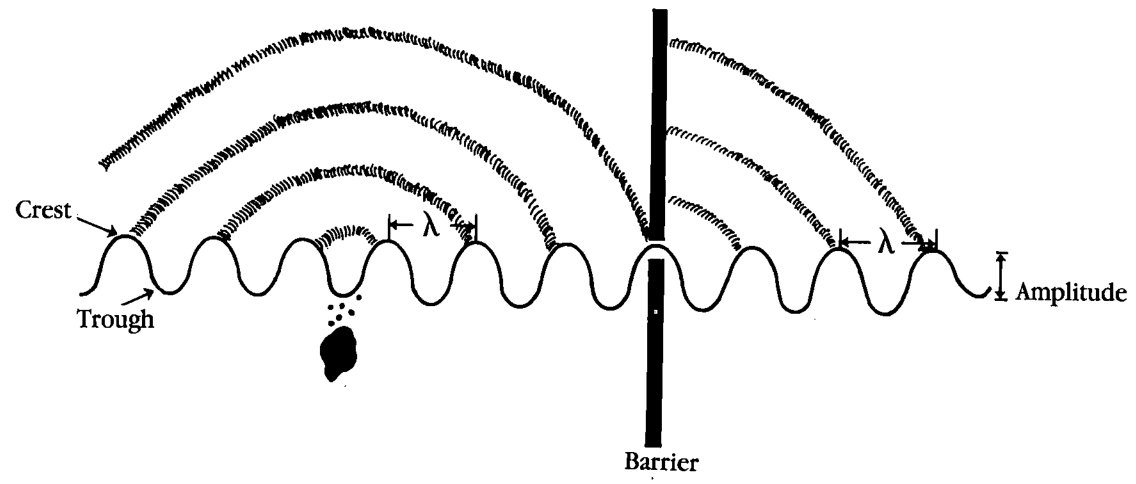

1-4 Waves created in water by a dropped stone are analogous light waves.

The waves that water makes are an adequate two-dimensional representation of light waves. This type of wave is called a *sine wave.* The bottom of FIG. 1-5 shows a typical sine wave, which increments in degrees. The measurement or increment in degrees can be used to show the phase relationship between two light waves. For instance, two waves in perfect phase would be 0 degrees apart.

Two light waves passing through the same point in space can be algebraically added together. This is illustrated in FIG. 1-5. Figure 1-5a shows two waves that are in phase, which mean crests fall upon crests and troughs fall upon troughs. When summed together, they create a wave with the added amplitudes of both waves. This is called *constructive interference.* Figure 1-5b

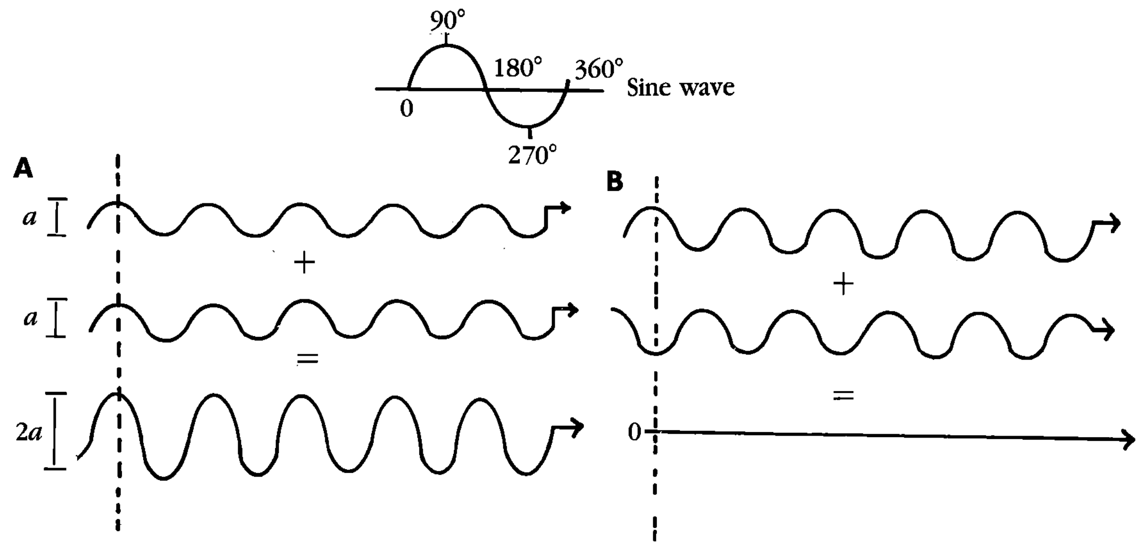

1-5 Interference of light, constructive and destructive.

details two waves that are 180 degrees or ½λ (λ⁄2) out of phase. Here, crests fall upon troughs. The algebraic sum of these two waves equals zero, an example of *destructive interference.*

Interference is not always completely destructive or constructive but varies in proportion to the amount one wave's phase is shifted in relation to the other. In FIG. 1-5a, if one of the waves shifts a full 360 degrees (one λ) in either direction, crests would still fall upon crests and troughs upon troughs. Essentially, the picture would appear identical. Summing the two waves together would be identical also, resulting in constructive interference. An inference from this information is that any two similar waveforms that are shifted a whole number of wavelengths apart (*n* times λ) produce constructive interference.

Inasmuch as full-wave shifts produce constructive interference, it's also true that two similar waveforms shifted ½λ (λ⁄2 or 180 degrees) always produce destructive interference.

In the early 1800s, Thomas Young performed a demonstration with light waves that is analogous to our water wave and interference illustrations. In his demonstration, Young used double slits for holes in the barrier. To make FIG. 1-6 clearer, the light waves impacting on the double slits are shown as plane waves. Huygens' principle states that each slit emanates its own wave. The waves emanating from the slits are spherical and undergo constructive and destructive interference as shown on the screen. The bright areas represent

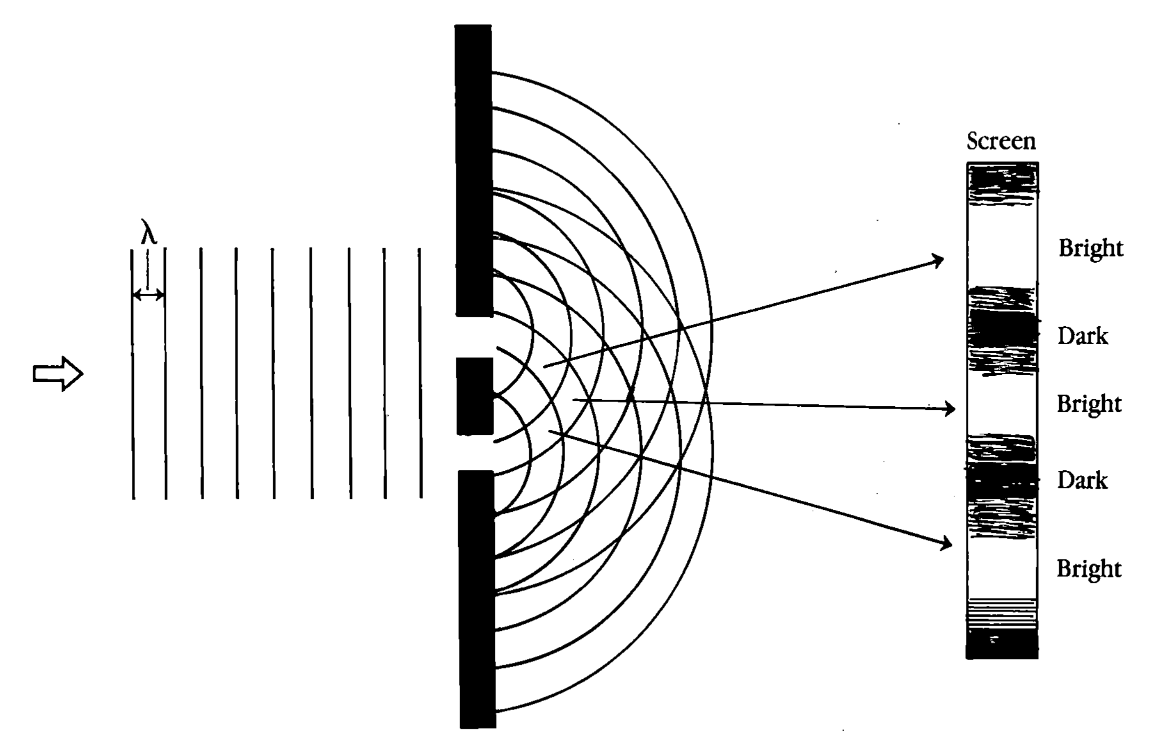

1-6 Thomas Young's double-slit demonstration.

constructive interference. The points to remember in this illustration are:

- Each slit behaves as a separate light source.
- Because each of the waves emanating from the double slit originated with the same wave impacting the barrier, the emanating waves are in phase and are identical in wavelength with one another.
- The interference pattern created by the two emanating waves can be observed on a screen or recorded on film.

This basic information provides the foundation to construct a simple hologram. Chapter 11, "Light," fully explains these concepts.

A simple hologram

Instead of using double slits for the two light sources, split a single beam of light emitted from a laser. The laser light is highly monochromatic (single frequency, single color) and it is in phase, two qualities that are required to record holograms. One beam illuminates the object. The light reflected off the object is illustrated as a series of spherical waves (FIG. 1-7a) that are equivalent to one source. In holography, this beam is identified as the *object beam* because it illuminates the object. The second light source is the other part of the split beam, shown as a series of plane waves traveling at an angle to the reflected waves from the object. In holography, this is called the *reference beam.* A photographic plate is positioned to record the *interference pattern* created by these two beams.

After making the exposure and developing the film, the interference pattern recorded on the film is a hologram. If you look at this film under ordinary white light, nothing resembling the image would be present on the film. It would appear as a mottled, grey-colored piece of film. If the film is illuminated with just the reference beam from the laser, the interference pattern recorded in the film reconstructs the original image (see FIG. 1-7b). The reconstructed image will be three dimensional and contain the vertical and horizontal parallax characteristics discussed earlier. It isn't necessary to use a laser to illuminate the hologram for viewing; a simpler monochromatic light source can be used for this purpose (a well-filtered white light).

How it works The *fringes* created by the interference of the two beams are recorded in the film emulsion. In the geometric model, the fringes are thought of as partially reflecting hyperbolic (curved) surfaces. So when light is shone on the emulsion, the interference pattern behaves like thousands of tiny mirrors, perfectly arranged in the emulsion, reflecting light to reproduce the image of the object. Not all of the incident light is reflected; some is transmitted through the plate, and some is absorbed.

GEOMETRIC MODEL

To get a better handle on the geometric model and to see how it operates for the popular white-light reflection hologram, let's take a small step backwards. For the time being, assume that each light source is a pebble dropped into water. Figure 1-8 illustrates the waves emanating from each source, labeled A

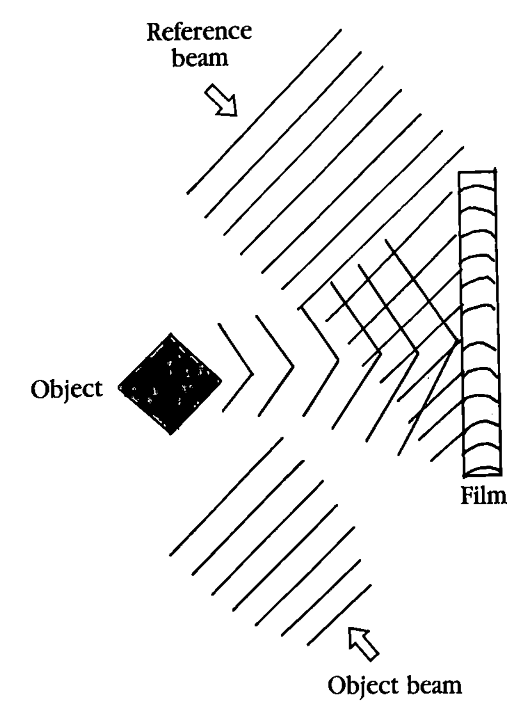

A. Recording interference pattern in film

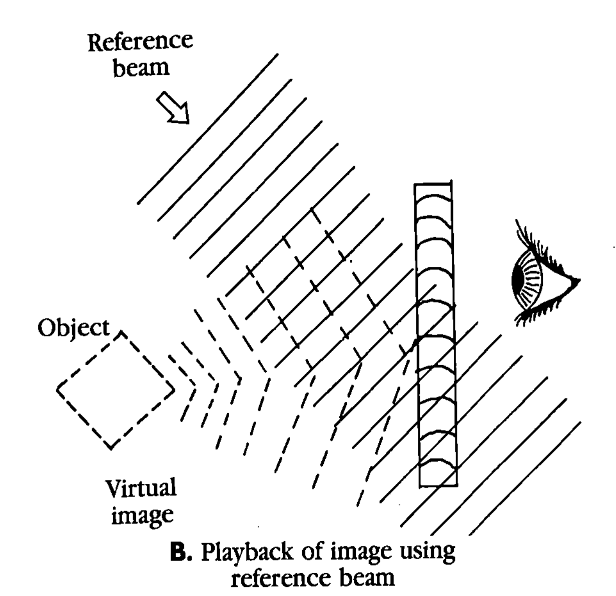

B. Playback of image using reference beam

1-7 Interference recording and playback transmission hologram.

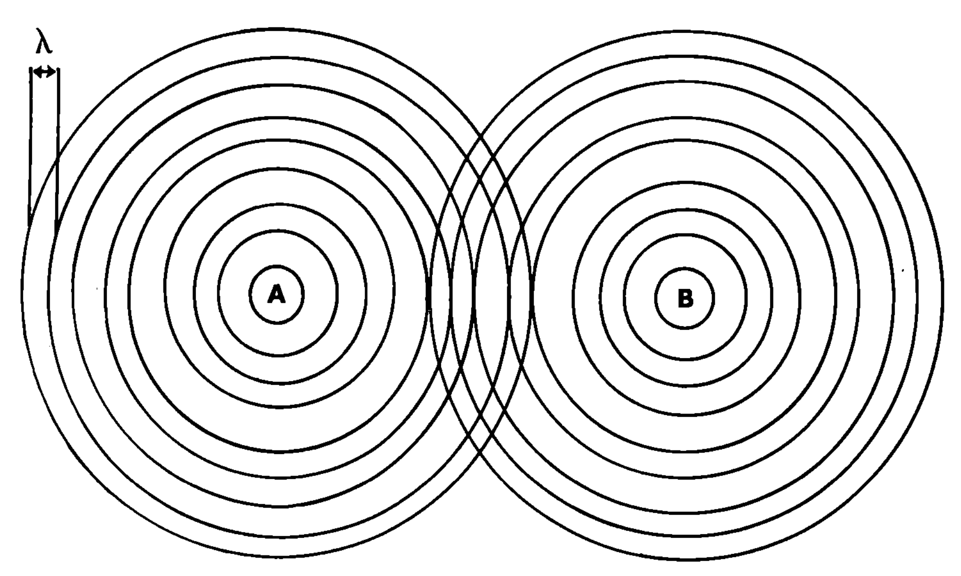

1-8 Wavefront and interference pattern generated by two light sources.

and B. Interference exists where the waves cross one another. Imagine each circle line in the illustration identifies the crest of the propagating wave. Hence, where two crests meet, there is constructive interference.

It isn't quite obvious from this illustration, but the two sources produce hyperbolic (curved) lines of constructive and destructive interference. To simplify the drawing and illustrate the point FIG. 1-9 shows only two waves, one

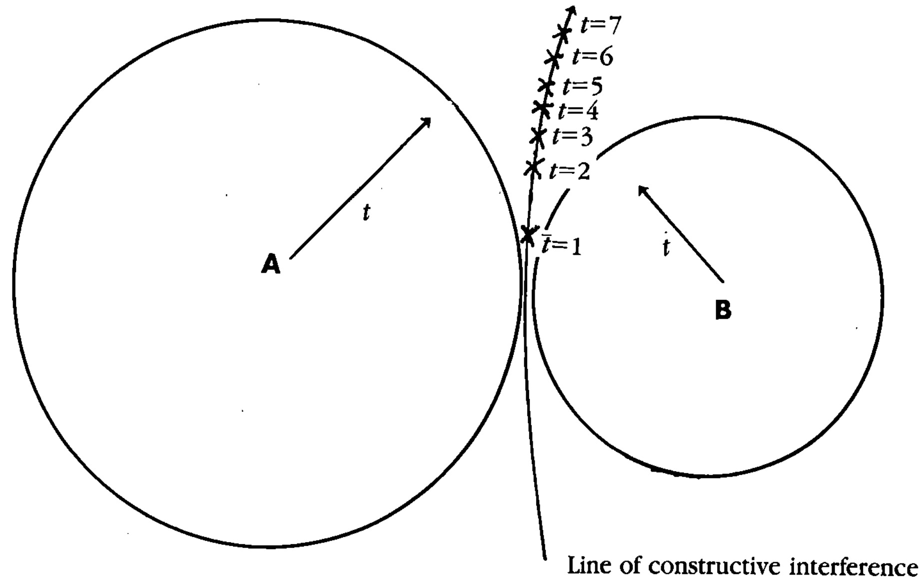

1-9 Generation of hyperbolic lines of constructive interference by two advancing waves over time.

from each source. The waves generate a hyperbolic line of constructive interference as the waves propagate over time (t).

Referring back to the original depiction of the two sources (FIG. 1-8), remove the concentric circles identifying the wave crests, and in their place draw the line of constructive interference produced by the two sources that it would resemble FIG. 1-10. The constructive interference lines are the fringe lines that are recorded in the film emulsion when making a hologram. Be aware that the spacing of the light waves and fringes in the figure are greatly exaggerated for illustration. The actual spacing is on the order of light wavelengths and are quite tiny. The spacing is estimated to be approximately 1 μm (10^{-6} meter) apart. (Actual spacing is determined by the geometry of the holographic setup and type of hologram being shot.) At that fringe spacing, the photographic film would require a minimum of 1000 lines per millimeter resolution to record an interference pattern.

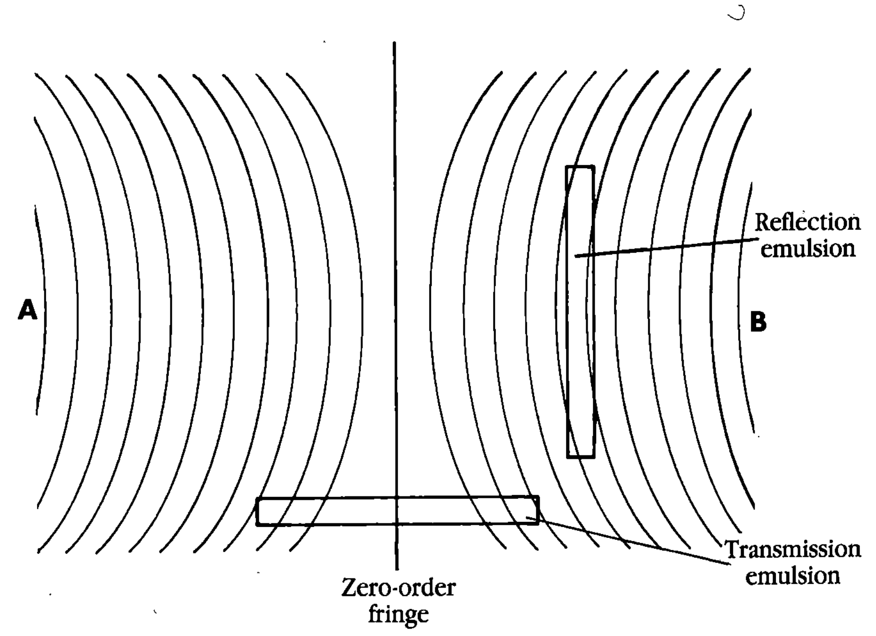

1-10 Hyperbolic lines of constructive interference generated by two light sources and placement of emulsions.

TRANSMISSION AND REFLECTION HOLOGRAMS

Where you place the film in the interference pattern determines the type of hologram generated. In FIG. 1-10, film emulsion #1 generates a *transmission*-type hologram, while film emulsion #2 generates a white-light *reflection* hologram. These two types are actually representative of families of holograms. During your study of holography, you will probably hear of many different types of holograms, but regardless of the name or type, it is always a derivative of either the reflection or transmission hologram.

A simple analysis of FIG. 1-10 shows that transmission holograms have both of the required light sources on the same side of the plate. The interference pattern created in these holograms is like a number of vertical slits perpendicular to the plane of the emulsion. On the other hand, a white-light reflection hologram has a light source on each side of the plate. The interference pattern created in this hologram is like a number of horizontal planes parallel to the plane of the emulsion.

In both cases, remember that the interference pattern is still hyperbolic, whether vertical or horizontal. These surfaces focus and reflect light similarly to a spherical mirror. The light so directed from the interference pattern creates the image.

Figure 1-11 illustrates a white-light reflection hologram. The advantage of this type of hologram over the transmission type is that it is viewable in ordinary white light. The light source is on the same side as the observer, hence it's name *white-light reflection.* A point light source such as the sun or a

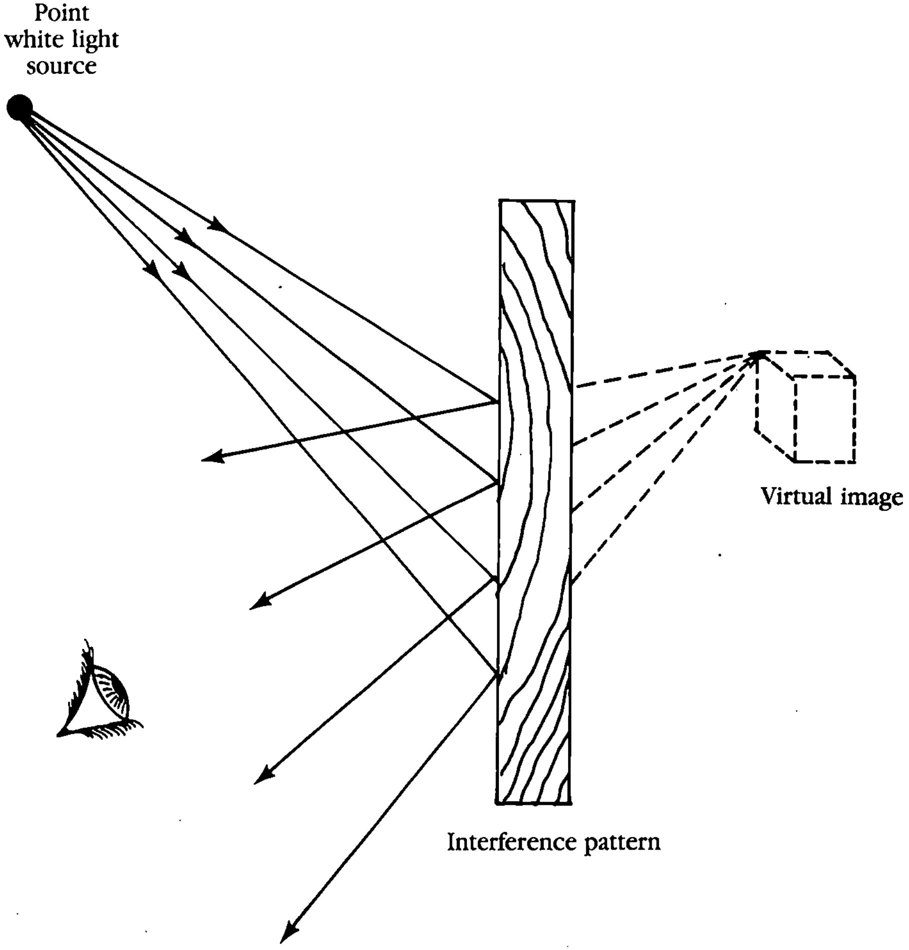

1-11 Image reconstruction, white light reflection hologram.

tungsten halogen lamp produces the brightest and clearest image. Diffused light sources like fluorescent lamps give poor images.

REAL AND VIRTUAL IMAGES

We have looked at the *virtual image* reconstruction without defining what a virtual image is. The most common example of a virtual image is the image that is reflected in a mirror (see FIG. 1-12). To the observer, the reflected rays that appear to come from the virtual image do not actually pass through the image. For this reason, the image is said to be *virtual.* The illustration also shows that the object and the virtual image are the same distance on either side of the mirror.

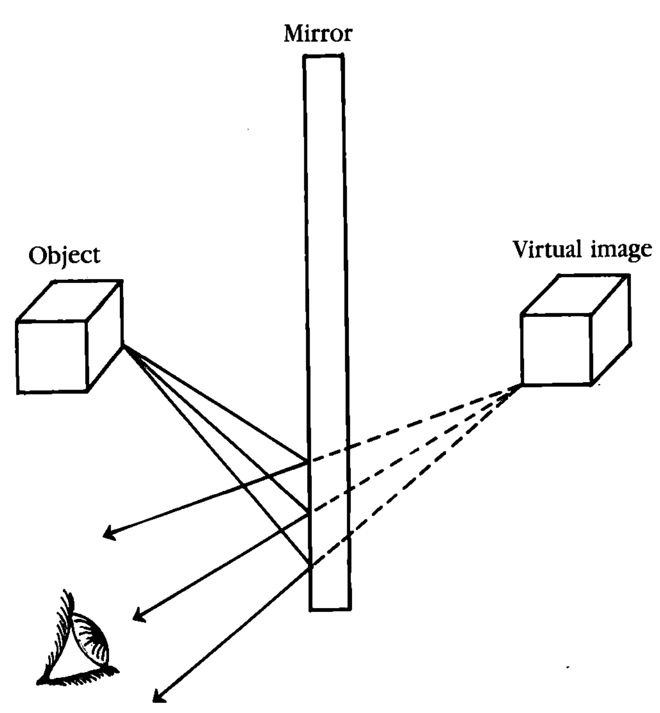

1-12 Virtual image, reflection in standard mirror.

The parallax and perspective of the virtual image in a hologram is observed to be correct. As you move your head from side to side or up and down, the three-dimensional scene changes in proper perspective as if you were observing the real physical scene. The virtual image is said to be an *orthoscopic* image, meaning a true image.

In contrast to the virtual image, a *real image* passes light through the image, making it possible to project the image onto a screen or onto film, as in a camera. Figure 1-13 illustrates a simple lens projecting a real image.

Real image in transmission hologram The transmission hologram projects a real image as well as containing the virtual image. Figure 1-14 illustrates

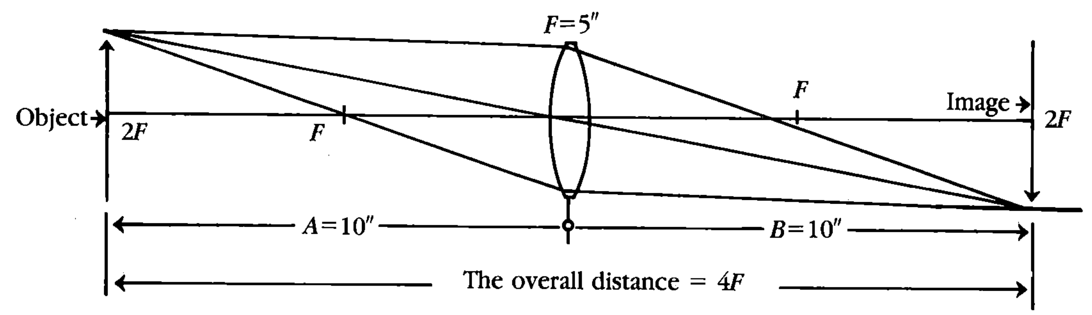

1-13 Real image projection by lens. In this particular case, image size is the same as the object size, which is useful in focus image/Fourier holograms.

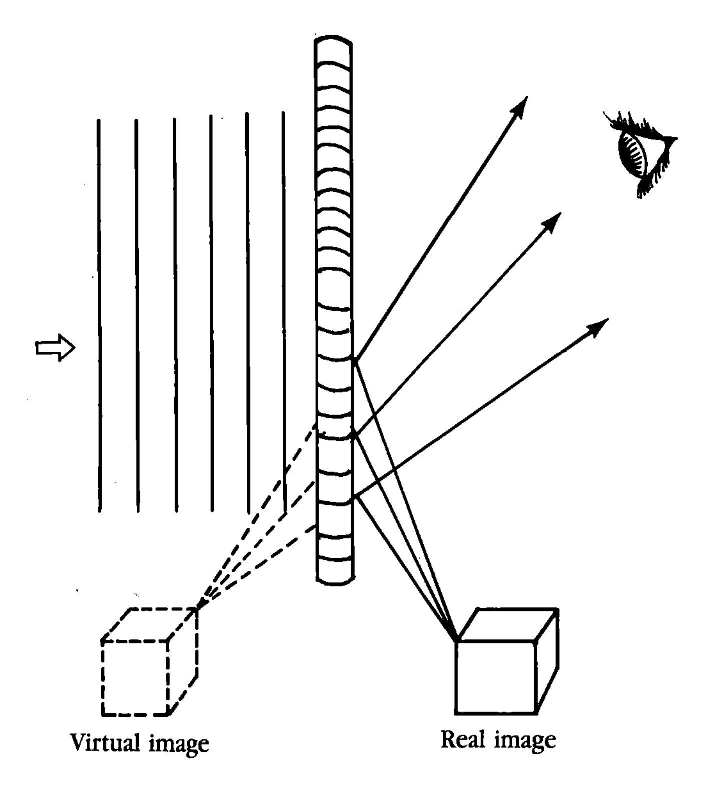

1-14 Real and virtual image reconstruction in hologram.

the real-image projection from the hologram. The real image in this configuration is quite dim. A better way to see the real image is by flipping the hologram around (180 degrees) as in FIG. 1-15. This essentially makes the real image bright and the virtual image dim.

The real image has some peculiar properties. The perspective is reversed. Parts of the image that should appear at the rear are instead in the front and vice versa. As you move your head to the right, the image appears to rotate in proportion to your movement and you see more of the left side, not the right

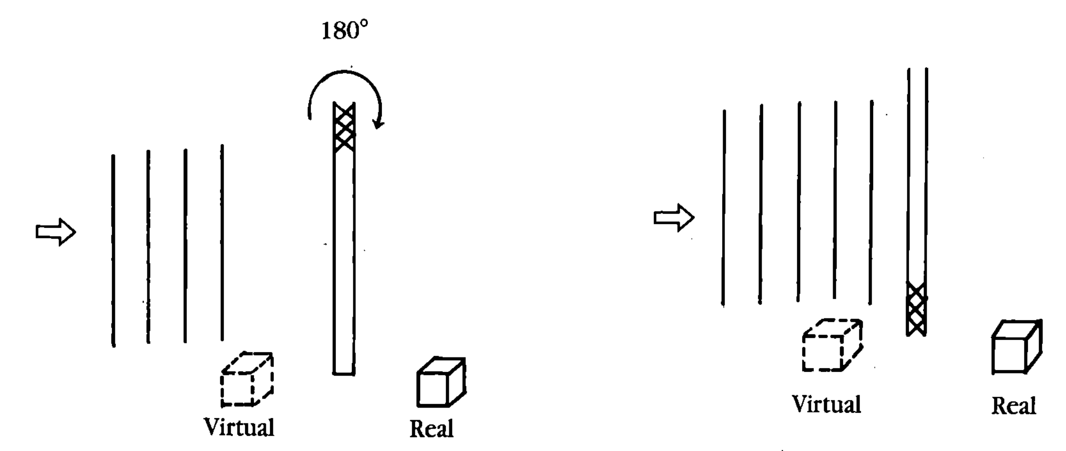

1-15 Flipping the hologram, which makes the real image brighter than the virtual image.

side. The brain perceives this paradoxical visual information as the image swinging around. The real image is said to be a *pseudoscopic* image, meaning a false image.

Real image in reflection hologram To see the real pseudoscopic image in a reflection hologram, simply flip the plate around (see FIG. 1-16).

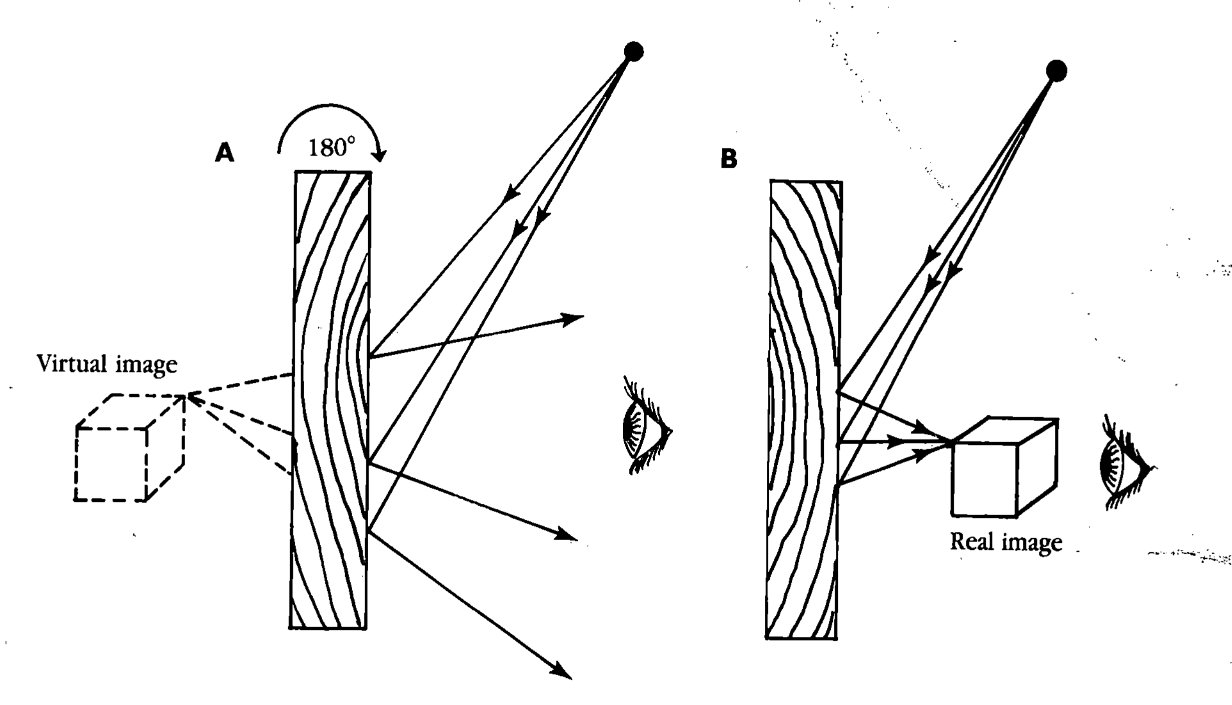

1-16 Real image reconstruction in reflection hologram. A shows the reflection hologram, and in B it is flipped.

Chapter 2

Isolation table

While the technology of holography is complex, the basic process and procedures for making holograms are not. Therefore, the chapters on the technological aspects of light and lasers are later in the book. Right now, let's begin putting the essential equipment together you need to start making holograms now.

The minimum equipment required are an isolation table, a laser, and a few optical components with mounts. The material supplies you need are film, and developing chemicals, and plastic trays.

The next several chapters in this book are interdependent on one another. Although this chapter covers building the isolation table, you need the laser and some optical components from the following chapters to test it.

WHY YOU NEED AN ISOLATION TABLE

Stability is essential to producing holograms. The isolation table stabilizes the components placed on it and isolates them from vibration. Since the surface of the isolation table supports all of the components and subject, the impact of vibration cannot be overemphasized. Vibration so slight that it is imperceivable to our senses can prevent the hologram from forming clearly.

How do you know if the table is doing its job? There is a simple test you can perform on the table to check it for vibration using a simple Michelson interferometer. Don't let the word *interferometer* throw you; you'll discover the apparatus is quite simple to put together and should only take a few minutes of your time.

There are a number of advantages to the table explained in this chapter. Most books on holography detail using sand tables. Sand tables have been the main workhorse for many holographers over the years. For others (like myself), it was a major impediment. After all, how many people can afford to keep a 500- to 2000-pound monstrosity in their home? It certainly isn't portable, and you can't put it away when you're done with it. For apartment dwellers, it's totally out of the question. Plus, a sand table requires considerably more work and money to put together than our simple table.

This table weighs about 20 pounds and costs approximately $30. It's cheap, portable, and can be set up or stored in a few minutes.

ISOLATION TABLE CONSTRUCTION

A list of materials needed to build the isolation table follows. See FIG. 2-1.

- One 8-foot piece of good, straight lumber that is 12 inches wide and 1½ inches thick (cut in half to 4-foot sections)
- Two small inner tubes from compact car tires (or equivalent)
- Two 1-foot-square pieces of carpet
- One steel plate, approximately 1-foot square and 1⁄16 inch thick

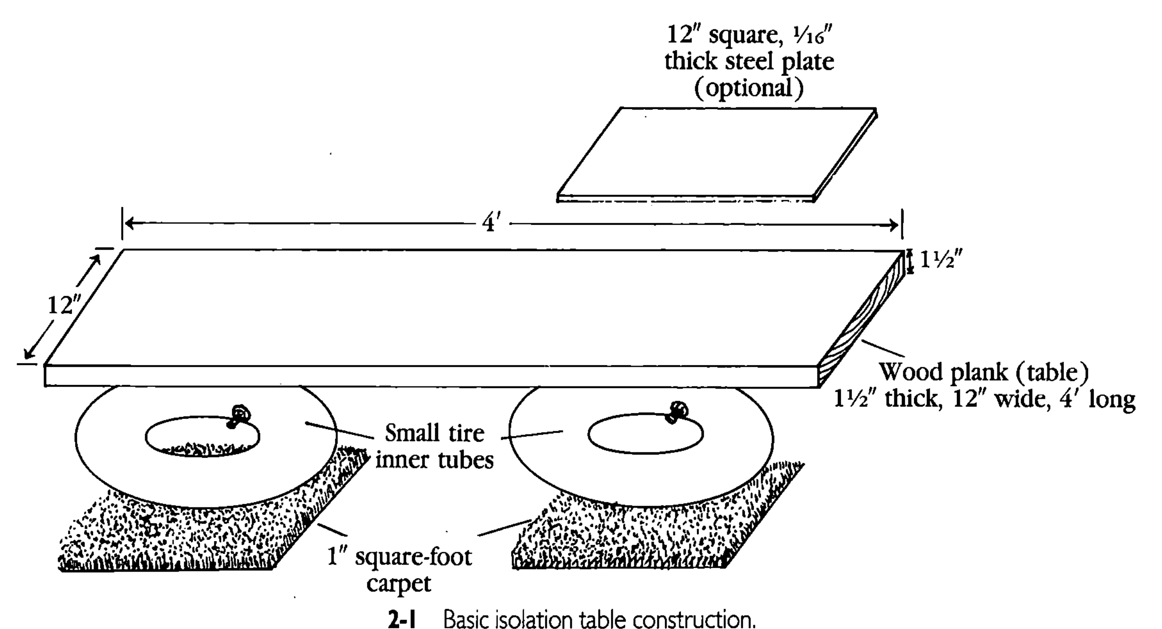

2-1 Basic isolation table construction.

Optional materials for greater stability (see FIG. 2-2) include:

- Two 1-foot-square, ⅔-inch-thick plywood
- Two 1-foot-square carpet
- Six 1½-inch-square sorbathane pads or balls

Cut the 8-foot piece of lumber in half because only one half of the lumber is used for this simple table. The other half is used when constructing the larger, more advanced table.

The inner tubes can be from any small tire. I found small inner tubes in a construction supply store. They were two small tires from a forklift. You could use a small inner tube from a compact car tire. Try to get an inner tube with a 18-inch or smaller outside diameter. Inflate the tubes until they are full, but still very soft.

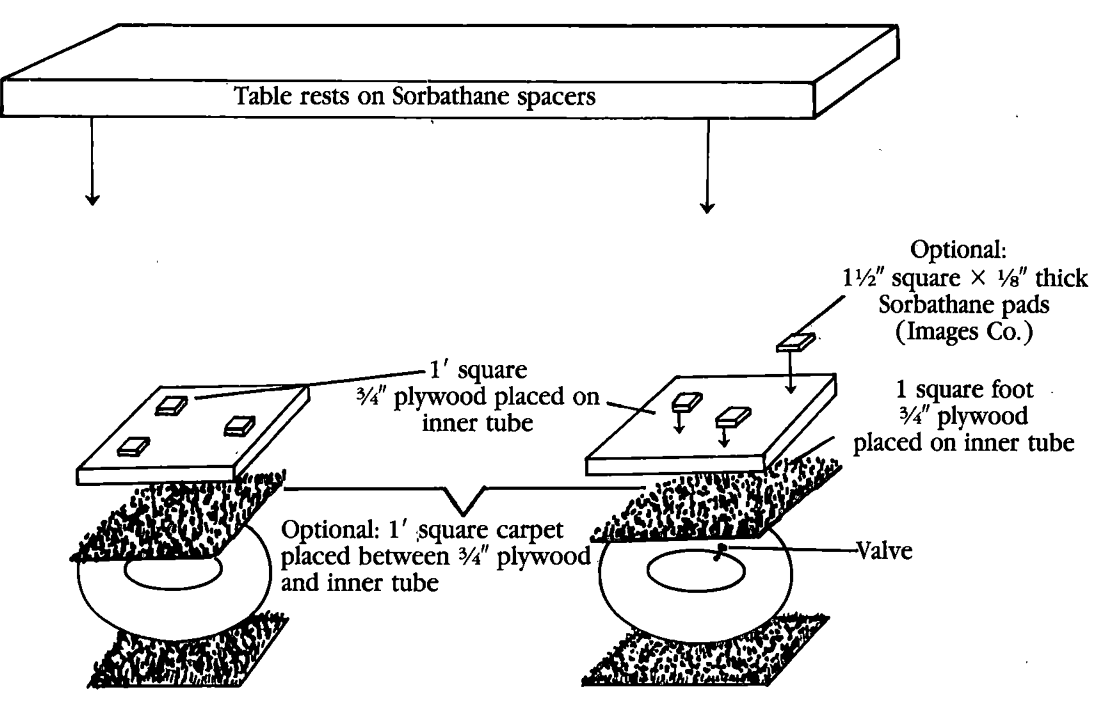

2-2 Improved table with optional materials for greater stability.

The carpet can be used remnants available for a nominal cost at a local carpet store. Or, cut two 1-foot-square pieces from a piece you happen to have around the house.

The steel plate really isn't mandatory, but it makes subsequent setups so much easier that you should really try to get one. Various components in these setups have magnets mounted to them, so it's easier to remove and replace these items from a steel plate than to use more permanent means that you might want to change or move later. The size and thickness isn't critical, but don't get anything thin enough that it will flex when something is placed on it.

TABLE LOCATION

Table location is important because some places are noisier than others, where *noise* refers to vibration. In particular, low-frequency vibration like that caused by passing cars and trucks or trains, is the most damaging. It's best to locate the table in the quietest area possible. If you live in a house, the ground floor or basement is best. A garage can also be a good work area. Usually, any place with a concrete floor will work well. If you live in an apartment building with wood floors as I do, it is still possible to shoot holograms, but you have to be more careful. In any case, don't lean the table against a wall or let any part of the table come in contact with the walls of the room, because this can add to the vibration.

The second consideration in picking a table location is light. You want to block it, stop it, or eliminate it altogether. The table area or room must be able to be completely darkened. Don't worry about working in complete darkness; you will use a green safelight. When shooting a hologram, the unexposed film plate is placed on the table for exposure. Therefore, any unwanted light will expose the film and fog it. An easy way to make a room light-tight is to cover any window with opaque plastic and tape it into position. To check the room, darken it and allow your eyes to become accustomed to the darkness. Any light leaks should then become visible. In addition, if you make it a habit of shooting holograms after sunset, this can add a measure of insurance.

The ASA rating (film speed) of the holographic film (discussed in the following chapter) is quite low, about 0.01 ASA. This low ASA is more immune to light leaks than regular film, but even so, it's best to get the room as dark as possible to be safe. As stated previously, don't become concerned about working in complete darkness; you won't have to because you'll be working with a green safelight. The safelight details and construction can be found in chapter 5, "Film."

Once you have decided where you are going to place the table, check the area for vibration. Set up the isolation table as shown in FIG. 2-1 in preparation to build an interferometer on it. You need a laser and four optical components. The optical components are one lens (−8 mm), two front-surface transfer mirrors, and a beam splitter. (Instructions for constructing the components are in chapter 4, "Optical mounts.") The interferometer can graphically display any vibration in the area or on the table.

MICHELSON INTERFEROMETER

Assuming you have the components necessary and the table is set up, this is the procedure to build a Michelson interferometer.

- Arrange the laser and one front-surface transfer mirror as shown in FIG. 2-3a. The front-surface mirror on the magnetic assembly permits horizontal, vertical, and tilt adjustments. Reflect the beam back to the laser so that the reflected beam is in close proximity to the aperture of the emitted beam.
- Place the beam splitter at a 45-degree angle, as shown in FIG. 2-3b so that the secondary beam produced is at a 90-degree angle to the first.
- Place second front-surface transfer mirror assembly as shown in FIG. 2-3c. Place the viewing screen opposite the second mirror assembly. The viewing screen can be any white surface, for example stiff paper or cardboard. Adjust the second mirror assembly to reflect the secondary beam back through beam splitter onto the viewing screen. The screen should have two red dots on it. Keep adjusting the secondary mirror so that the two red spots on the screen come together and merge into one spot.
- Place the 8mm diverging lens as shown in FIG. 2-3d. The diverging lens is also on a simple magnetic assembly that permits horizontal, vertical, and tilt adjustments to the lens. The diverging lens spreads the spot, making it possible to see alternate light-and-dark line patterns. After the lens is put in place, remove your hands and don't touch the table. Allow a

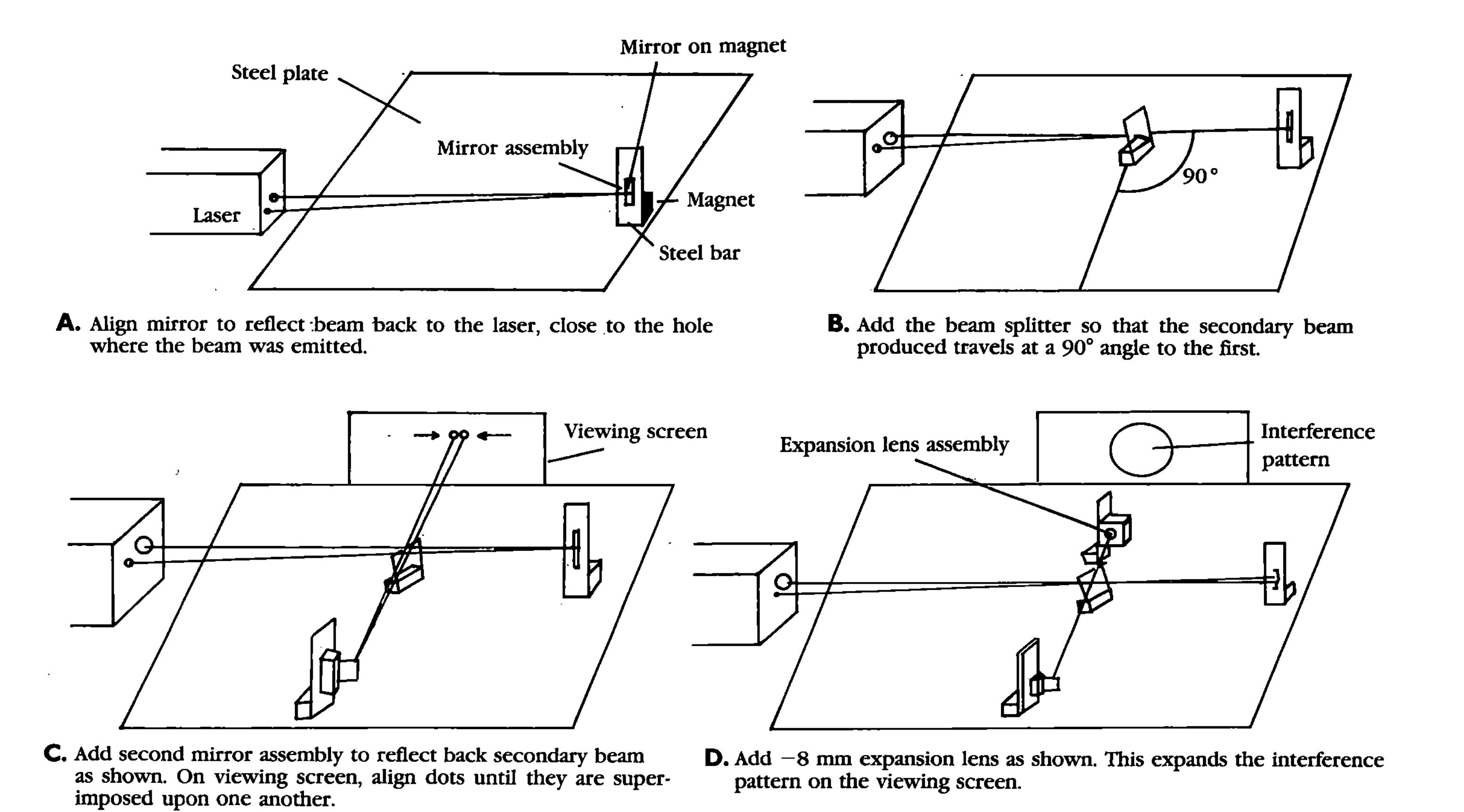

A. Align mirror to reflect beam back to the laser, close to the hole where the beam was emitted.

B. Add the beam splitter so that the secondary beam produced travels at a 90° angle to the first.

C. Add second mirror assembly to reflect back secondary beam as shown. On viewing screen, align dots until they are superimposed upon one another.

D. Add −8 mm expansion lens as shown. This expands the interference pattern on the viewing screen.

2-3 Setting up the Michelson interferometer.

minute or so for the table to stabilize (stop vibrating) and the fringe lines to appear. Figure 2-4 is a photo of the fringe lines made from the interferometer.

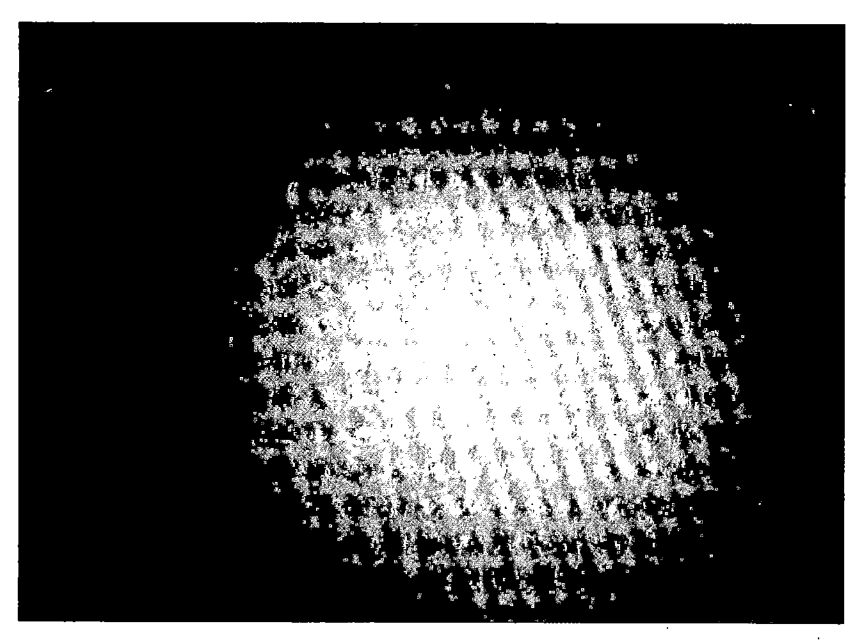

2-4 Photo of fringes line created by interferometer.

Troubleshooting

If fringes don't appear, try carefully readjusting the secondary mirror. Depending on the adjustment, you might have a glob of bright light in the center with the fringes only visible on the perimeter of the light glob. Try tapping the table lightly to see if fringes on the perimeter disappear and then reappear in a minute or so.

Try darkening the room; this might make the fringes visible if they are being washed out by the room light.

If you are playing any music from a stereo or radio, turn it off. The sound could be vibrating the components or the table.

Figure 2-5 is an overview of the interferometer. Adjust the mirror distances distances A and B from the beam splitter to be equal to maximized fringe contrasts.

Although the magnetic mounting system that I designed is simple, it is nonetheless quite stable. Because its design is so simple, you might think that just about any mounting can be used, but not so! If you substituted different mountings for your optical components, it could be the mountings that are unstable, not the area or table.

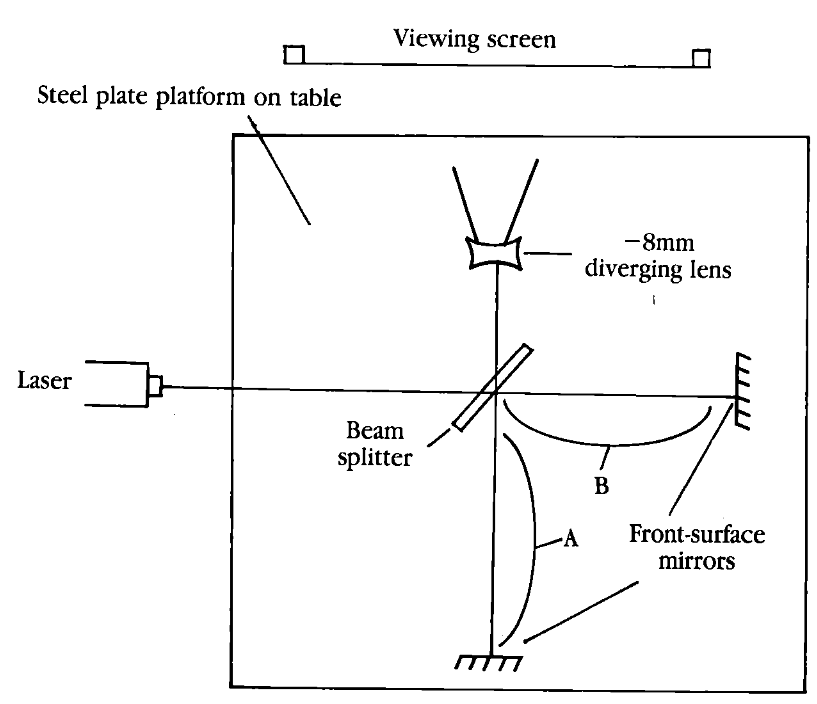

2-5 Top view of interferometer setup.

If you have tried everything and you still can't see any fringes, but the optional materials explained in FIG. 2-2 and put them into the table to provide greater stability. The sorbathane pads are an extremely resilient rubber product that has a tremendous dampening ability.

If the fringes still don't appear after adding the optional stability material, you might have to try another location.

TESTING

If the fringes are visible, congratulations. My feeling is that setting up an interferometer is more difficult than actually shooting a simple hologram. The difficulty lies not in the mechanics involved but in not knowing what you are looking for the first time around.

To do some tests: first tap the table. Notice the fringes disappear immediately. Check the elapse time from when you tap the table until the fringes reappear. This is the table's *relaxation time.* Because the table isn't massive, the relaxation time is usually quite short. This relaxation time is the minimum amount of time you must wait after finishing a setup on the table. Usually, the last thing you do is place and position the holographic plate on the table. In some cases where the area is quite stable or with using the sorbathane pads, the fringes reappear almost instantly. Don't let that throw you off. That just means you have an excellent area to work in.

Bang or stomp around on the floor around the table, and check the impact this has on the fringe stability and the subsequent relaxation time. If you walk around the table after it is set up, you must take this time into consideration before exposing the film.

Blow some air on the interferometer setup. This should also cause the fringes to move or disappear. If so, a drafty area would have a negative impact on your holograms. (Some professional holographers put curtains or cards around the entire table to prevent any movement of air from degrading the image.)

When you first shoot a hologram, you will be using a shutter card to block and unblock the laser and thus control the exposure. Check the impact that lifting a shutter card off the table has on the fringes. When you lift the card off the table, the fringes will move. Therefore, when shooting a hologram, first lift the card off the table, and then hold it stationary and above the table for a minute to allow any vibration to dampen out before moving the card out of the path of the laser to make the exposure.

Drift

If the fringes do not stay stationary or slowly drift, make sure the table isn't touching anything, including the walls. Another possible cause could be that the inner tubes are a little overinflated. Try removing some air from the tubes and see if that helps. You could also add the optional materials for added stability (FIG. 2-2). Or, if you have moved the table from another place, the table temperature could still be adjusting to the new location. If the fringes drift when shooting holograms, you will get either no hologram or a poor one.

Equipment interference

The interferometer can be used to check the impact any piece of equipment has on the table, such as a piece of equipment that moves (an electronic shutter or stepper motor assembly). This book has construction plans for a electronic shutter. If you build an electronic shutter, place it on the table along with the interferometer. Turn the shutter on and off a few times. Check for any impact on the fringes as the shutter goes on and off. (The shutter is designed to have no impact on the fringes.) Another important test that can be conducted is checking the coherence length of the laser.

CHECKING THE COHERENCE LENGTH OF YOUR LASER

Checking the coherence length of your laser is not mandatory or a prerequisite to producing holograms. It is a technique to determine the maximum size or depth capable of being produced with the particular laser you are using. None of the holograms described in this book approach the limits imposed by the coherence length of the laser. Consider this an advanced technique. Do not allow yourself to get bogged down checking the coherence length of the laser because it isn't absolutely necessary. Just remember it's described here in case you have a need for it.

The coherence length of the laser is an important factor. It determines the maximum depth that can be holographed. Typically, HeNe lasers have a coherence length of about 8 inches. The coherence length of the HeNe laser is explained in greater detail in chapter 3, "Lasers". Suffice it to say here that lasers are not perfectly monochromatic. HeNe lasers emit more than one frequency of light, but the frequencies emitted are extremely close, varying by only about 0.00065 nm.

As these two emitted beams travel over distance, the phase difference between them varies between constructive and destructive interference, depending upon the exact distance. Checking the coherence length of the laser measures the distance from where the frequencies are in phase (constructive interference) until they go out of phase (destructive interference).

The impact this length has in holography is that when a laser beam is split and recombined to produce a hologram, if the path differences vary by an amount that produces destructive interference at the holographic plate, the hologram will not come out. This is why when shooting split-beam holograms, the distances of each optical split beam path to the film must match. This way, the production of the interference pattern is ensured. With the optical paths so matched, the coherence length now determines the depth of the viewable scene in the hologram.

The procedure for measuring the coherence length follows:

- It is assumed you have set up the Michelson interferometer and the fringes are visible. Refer to FIG. 2-5, the top view of the interferometer. Move both front-surface mirrors close to the beam splitter. Make sure both distances A and B are equal. This should produce high-contrast fringes on the viewing screen.
- Move one of the two mirrors back ¼ inch at a time. Observe the fringe pattern each time. You should see that as one mirror is moved further and further away, there is a loss of contrast in the fringe pattern.
- When you have moved the mirror to the point where the fringes cannot be seen clearly, the distance traveled is one-fourth the coherence length of the laser. To get an even more accurate measurement, if you still have sufficient room to move the mirror back, continue to move the mirror back until the fringe contrast peaks. The distance traveled is now one-half the coherence length of the laser. For a 5 mW HeNe laser, this distance is about 4 inches, which gives you a coherence length of 8 inches.

Chapter 3

Introduction to lasers

The next piece of equipment to consider is a laser. There are many different kinds of lasers available. The main laser families are gas, solid state, and semiconductor. Each family of lasers subdivides into various types. For instance, gas lasers include helium-neon, carbon dioxide, argon, helium-cadmium, and metal vapor. What distinguishes these lasers from one another, aside from the working gases used, is power output and wavelength.

For the holographer, power output is inversely related to exposure time. The higher the power output, the shorter the exposure time. Frequency or wavelength relates to color of the laser light and determines the type of emulsion (film) you should use with it.

The helium neon (HeNe) laser produces a red beam that centers around a wavelength of 632.8 nanometers. Since the HeNe laser has been the workhorse laser of holography community and industry for many years, there are a number of companies that sell holographic film for it. This is the type of laser that is used in this book.

Other types of lasers are also used in holography. Lasers of various frequencies are used (in combination with other lasers) to generate color holograms. Pulsed ruby lasers are used for pulse portraits. Some photo-resist and photo-polymers used in commercial preparation of mass-produced holograms are insensitive to red HeNe laser and require a different laser to be exposed properly (most notably argon). Other emulsions have a higher efficiency (make brighter holograms) than HeNe emulsions.

Pulsed ruby lasers are used to make pulse portraits of people, flowers, and animals. Because of the problems with vibration discussed earlier in reference to the isolation table, it's impossible to make a hologram of a person using a HeNe laser. No one alive could possibly keep still long enough for stationary interference fringes to be formed and recorded. It therefore becomes necessary to use a pulsed laser that can deliver the required energy to expose the plate in a very brief period of time. Because of the very short exposure time, vibration or movement becomes less critical.

More complete information on lasers is in chapter 12, "Laser Technology."

CRITERIA FOR THE LASER

There are certain criteria the HeNe laser must meet for it to be suitable for holography. First, it must be a *continuous wave* type, abbreviated CW. This simply means that it produces a continuous beam rather than pulsed light. Second, it must operate in TEM_{00} mode. TEM stands for *t*ransverse, *e*lectric, and *ma*gnetic modes respectively. Suffice to say that TEM_{00} mode lasers provide the smoothest distribution of laser energy across the beam's width. A fuller treatment of TEM modes is provided in chapter 12.

The next factor isn't essential, but it is useful. Try to obtain a linear polarized laser. It might appear that you get more milliwatts per dollar for an unpolarized laser, but the quality of the holograms won't be as good. This leads to the final consideration: power output. Power output of HeNe lasers is measured in milliwatts (10^{-3} watt). That's 1⁄1000 of 1 watt similar to 1⁄1000 of a standard electric light watt. The greater the power output of your laser, the shorter the exposure time. The shorter the exposure time, the less likely vibration or *creep* will ruin your hologram. Anything between 1 milliwatt and 5 milliwatts should work fine, but I suggest you try to get a laser on the high-power side.

LASER SAFETY

The most dangerous part of a laser is the high-voltage power supply. A 5-milliwatt laser is still considered a low-power laser and is relatively safe. Notice the operative word *relatively* in the last sentence. A 5 mW laser is completely skin safe, so you don't have to worry about any exposure to the skin. The eyes, however, are another matter. ANSI (American National Standards Institute) publishes a report of recommended safe eye exposure levels of laser light. ANSI uses the terminology MPE to denote *m*aximum *p*ermissible exposure. *Aversion response* is the approximate reaction time it takes a person to avoid (closing your eyelid and moving your head) a laser beam from hitting the eye. Aversion response time is 0.25 of a second. Because eye safety is so important, there is a more complete breakdown of the ANSI standard in the appendix A, "Laser Safety," but first read through this section for an overview.

An unspread laser beam emanating from any HeNe laser from 1 to 5 mW in power is well above the MPE aversion response. Hence, *under no circumstances should anyone look directly into the bore of an unspread laser beam.* This situation changes dramatically when the laser beam spreads. As an example, if the 5-milliwatt laser beam is spread with a lens to a diameter of 8 inches, it's considered more eye safe. Five milliwatts might not seem like a lot of power, but the effects of bright lights on the eyes are cumulative, so making a habit of carelessly shining a laser into your eyes only increases the chance of eye problems later in life.

LASER CLASSIFICATION

The U.S. government classifies lasers by their power output. The classification is established by the Center for Devices and Radiological Health (CDRH) of the

Food and Drug Administration. The requirements listed by the CDHR must be adhered to by manufacturers of laser systems, but these regulations don't directly impact users of lasers. Knowing the classifications for HeNe lasers can be useful when purchasing a laser. The classification is usually quoted in the laser specifications. These classifications are based on the output power of the laser that is accessible to human beings during normal operation.

Class I Output is below levels where biological hazards have been established.

Class II System may not exceed one milliwatt of output power and must contain a pilot light and beam shutter device.

Class IIIa Output power is between 1 and 5 milliwatts. The requirements for this class are similar to Class II with the exception of labeling. Most HeNe lasers produced are in this class.

Class IIIb Output power is between 5 and 500 milliwatts. Laser may not exceed 500 milliwatts of output power and must contain a pilot light, beam shutter, on/off key switch, provisions for remote operation, and a delay of several seconds from the time the switch is keyed to the time laser output commences. Skin and ocular hazards.

Class IV System exceeds power limits of Class IIIb and has safety features and requirements similar to Class IIIIb.

BUY OR BUILD?

My answer to the question of whether to build or buy is that I recommend building your own laser, mainly because it's cheaper that way, costing you one-fourth of the price of a new laser system. This substantially reduces your initial investment and puts holography in closer reach for those working with a limited budget.

New, 5-milliwatt laser prices are listed below (as of 1/90).

Manufacturer	Part #	Price
Newport	U-1305P	$750
Aerotech	LS5R	675
Aerotech	LS5P	735
Milles Griot	051HP851	625

The companies' addresses and phone numbers are in the Suppliers appendix. Edmund Scientific is a distributor for a few laser manufacturers.

Surplus laser systems

Many electronic surplus houses sell complete laser systems. "Complete" means power supply, laser tube, and housing. Typically, a surplus 3 mW HeNe laser costs about $200. The Suppliers appendix, lists a number of surplus houses that recently have listed surplus lasers for sale. Some "bargains" are much better than others, and some aren't bargains at all. It's up to you to determine. Along with the company listing is their most recently advertised laser. To see whether a particular laser is still available after this writing, you have to call or write. If the exact laser isn't available, most will offer an equivalent.

If you decide to go this way, keep in mind these particular points. Remember, any laser you purchase must match the criteria established previously in this chapter or the laser isn't suitable for holography. Most notable is the TEM_{00} mode requirement. If the supplier doesn't know the TEM mode of the laser, don't buy it. Check to see if the surplus laser is new or "pulled." The latter case means it was pulled from another piece of equipment. The disadvantage to a "pulled" unit is that the amount of hours on the laser is hard to determine. Laser tubes last between 10,000 and 20,000 hours, so a unit that has 5000 hours on it can still be a good buy. In any case, check the warranty offered with the laser and their return policy. When you receive it, start it up right away. Don't wait until you're ready to shoot holograms to check the laser out. Start it and let it run for a few hours. Check for overheating or any noticeable power output deviation in the beam's strength. If the laser you purchased is polarized, you can check the polarization with some polaroid material (see chapter 14). If you encounter any problems with the laser, contact the company you purchased it from, explain the problem you encountered, and return it for replacement or credit.

Many lasers on the surplus market are new but are manufacturer or customer rejects. Because a laser is a reject doesn't mean it's useless. It might only mean that the laser didn't meet a particular specification. Find out why the laser was rejected. If it still meets the criteria for making holograms, it can be used.

Surplus tubes New, polarized 5 mW laser tubes last approximately 20,000 hours and cost between $350 and $450. Pulled laser tubes available from surplus houses like Meredith Instruments and MWK Co. are considerably cheaper than new tubes and usually have thousands of hours left on them. Polarized 5 mw surplus tubes typically cost $100, and 2 mw tubes can be found for as cheap as $50. The power supply I designed can power any size laser tube from 1 to 5 mw and costs approximately $50 to build.

Surplus power supplies If you don't want to build a power supply, you have the option of buying a surplus power supply. Companies like Meredith Instruments, MWK, and Allegro also carry surplus power supplies in addition to laser tubes. A power supply for a 5 mW tube typically costs $80 to $110. If you decide to purchase the power supply, you would then just have to build housings for the laser tube and power supply. The cost saving would still be excellent compared to the cost of new units.

Building a laser

You can build a 5 mW laser for about $150 total. "Build" means just constructing the power supply. The heart of the laser, the HeNe tube, is a purchased item whose cost is included in the $150 estimate. The power supply I designed can power and size laser tube from 1 to 5 mW and costs about $50 to build. Chapter 14 has the schematics for building this supply and also contains information on mounting and housing the laser tube and connecting it to the power supply. Also included is information on wiring and housing a purchased power supply.

Chapter 4

Optical mounts

Most of the optical components—lenses and beam splitters for this project were purchased from Edmund Scientific. The magnets and steel plates used to make the mounts were purchased from Images Company.

The steel plates measure 4¼ × 1 × 1⁄16 inch thick. The plates can be assembled in a number of configurations using small bar magnets. I advise buying at least 10 of these plates to start with. Their usefulness on the table is versatile; they can be arranged in an almost infinite number of configurations. The most basic configuration looks like an upside-down T.

Putting the units together is simple. Figures 4-1 and 4-2 illustrate the assembly of two plates in the inverted T. The magnet at the bottom secures both plates together. The unit is more stable than it might appear.

THE LENS

The main purpose of the lens in holography is to spread the laser beam. This can be accomplished with either a *positive* or a *negative* lens. Figure 4-3a shows six basic lens shapes. The upper three are positive lenses.

With positive lenses, incident light is focused to a point at the focal length of the lens and spreads out from that point (see FIG. 4-3B). Typically, the focal length is measured in millimeters (mm) from the surface of the lens. If there are 25.4 mm to an inch, then a lens with a focal length of 12 mm would focus a beam approximately half an inch in front of the lens. Naturally, the shorter the focal, the more strongly the beam converges onto the focal point, which is directly proportional to how fast the beam spreads from the focal point. A lens with a 9 mm focal length spreads a laser beam to about a 4-inch diameter in a distance of about 2 feet.

The negative lens acts as if the focal point is behind the lens and spreads the beam out as if from that point. Because the focal point is behind the lens, the focal length is identified as a negative number, such as −8mm. Figure 4-3C

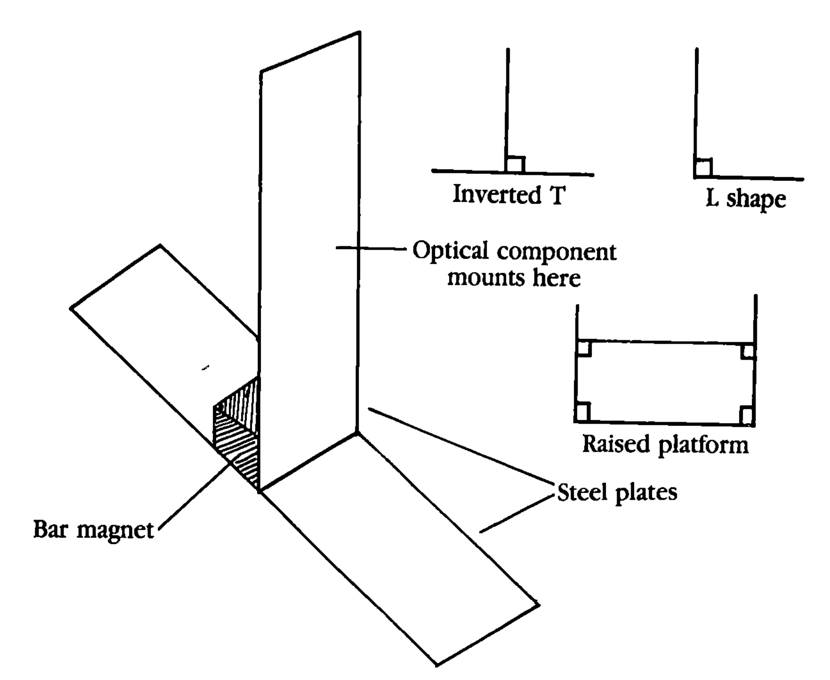

4-1 Diagram of basic optical stand for components.

4-2 Optical stand for components.

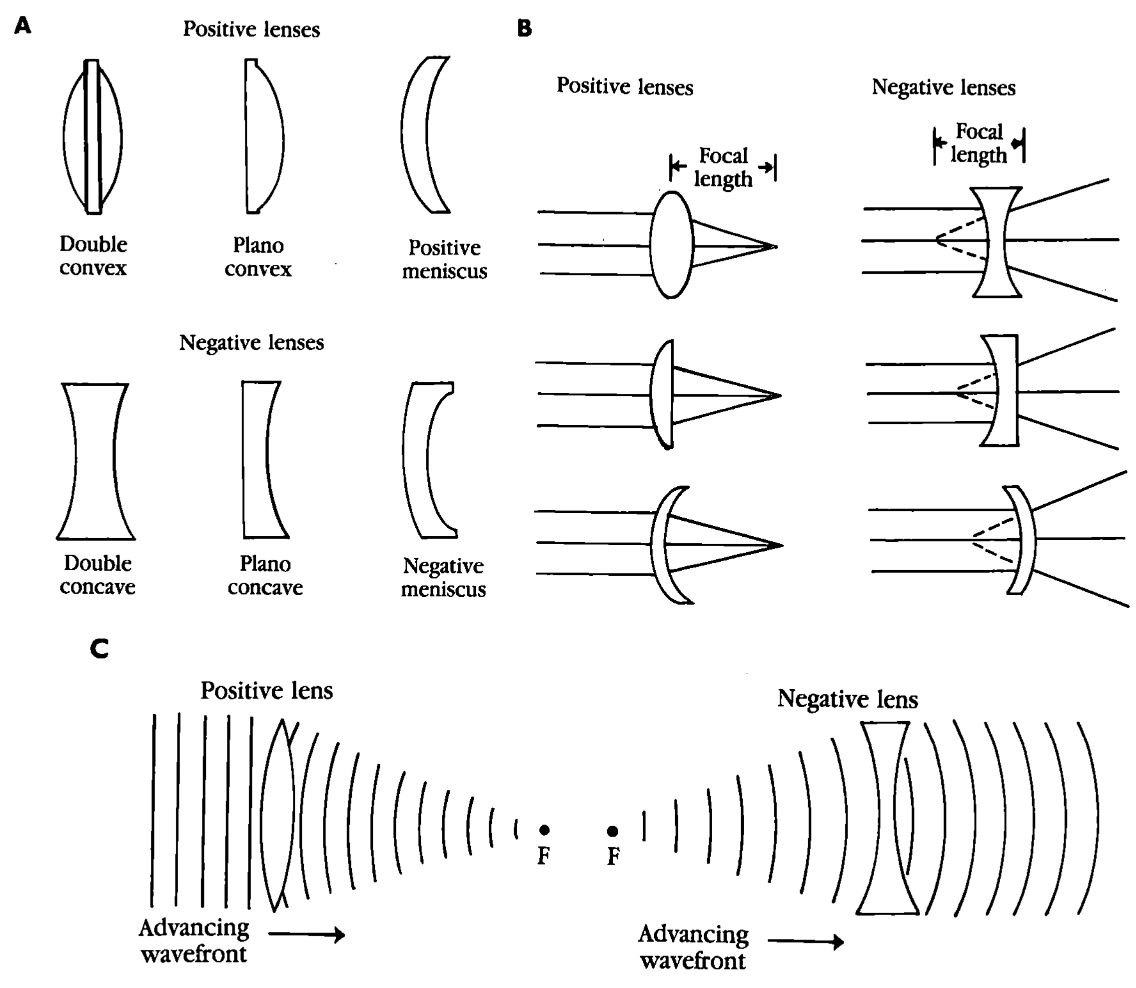

4-3 Operation of simple lenses.

illustrates the lens action on advancing plane waves. This is a typical type of illustration that demonstrates the coherence and phase of laser light.

It has been my experience that a negative lens can spread a laser beam with less added noise than a positive lens. Even so, I recommend that you purchase both types of lenses, because the positive lens can be used to make a simple spatial filter later on, which is something you can't do with a negative lens.

Microscope objectives

Many holographers use microscope objectives in place of simple lenses with the belief that a high-powered objective (60×) spreads the beam faster and much cleaner (less added noise) than the simple lens. It has been my experience that microscope objectives are not worth the additional monies they cost. Although it can spread a beam faster, I have seen a lot of noise coming from microscope objectives. You would be much better off using the small spherical mirrors (described later). If it is your desire to use a objective, you can still use the magnetic mounts system. Glue the barrel of the objective to a bar magnet and use it as any other component.

Securing optical component to magnet holder

The lens, positive or negative, is glued to the center of the rectangular magnet (see FIG. 4-4). The magnet has a hole in its center. The lens is glued so that its center lies upon the center of the hole in the magnet.

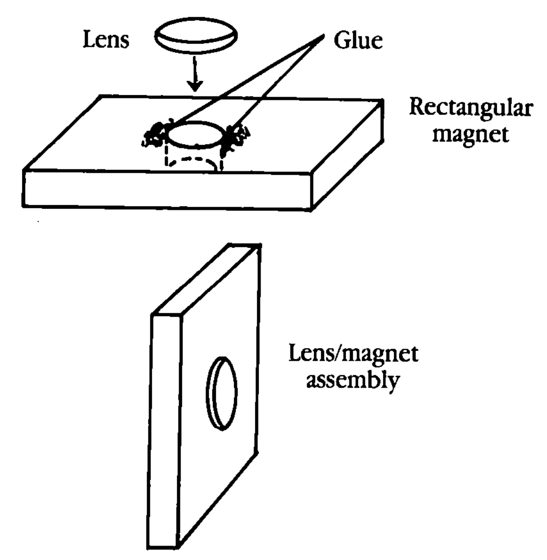

4-4 Mounting of lens onto rectangular magnet.

The magnet I used was purchased from Images Company (see Suppliers appendix). I recommend buying all magnets from Images Company. Although this type of magnet is available at a lower cost from Radio-Shack, it is not as powerful. A less powerful magnet can lead to problems; if the magnet should move slightly, it will redirect the beam and throw the lighting off. This becomes more critical in split-beam setups using the spherical mirror.

In reference to the glue, use a two-part epoxy, such as Devcon. Epoxy is available in most hardware stores. Mix the two parts according to the directions. When gluing lenses or other optical components, use only a very small amount of epoxy. In the case of gluing a lens to the rectangular magnet, only put a tiny amount on the edge of the hole in two places. That will be more than enough to hold the lens securely. If you apply too much glue, the lens will probably slide around and you end up with epoxy on the back surface of the lens. If you should end up covering the back surface of the lens with epoxy, take the lens off, wait for the epoxy to harden, and then scrape off the lens and the magnet with a razor blade and start over. It's important to keep the lenses as clean as possible. The cleaner the lens, the cleaner the laser beam will be traveling through the lens. Figure 4-5 is a photo of a −8mm negative lens glued

4-5 Negative lens (−8mm) mounted on magnet.

to a rectangular magnet. Figure 4-6 is a photo of a 9mm positive lens glued to a rectangular magnet.

When the epoxy hardens, the magnet/lens assembly is placed on the vertical plate of the inverted T assembly (see FIGS. 4-7 and 4-8). Vertical adjustments are made by moving the lens/magnet up or down. To adjust the angle, simply tilt lens/magnet to whatever degree is needed. Adjust horizontally and rotationally by moving the entire assembly left or right or rotating to whatever degree you want. There is essentially complete adjustment through all axes.

4-6 Small, positive lens mounted on magnet.

Front-surface mirrors

Mirrors are commonly used to redirect light around a holographic table setup. These are called *transfer mirrors.* Other uses for mirrors are as beam spreaders and collimators. For the time being, let's be concerned with small transfer mirrors and return to other mirrors and uses later.

In holography, it is necessary to use front-surface mirrors. Standard household mirrors are back-surfaced mirrors, meaning they have their reflective coating behind the glass (see FIG. 4-9). Front-surface mirrors, as their name implies, have their reflective coatings (aluminum or silver) on the front. As illustrated in the diagram, the back-surface mirror would create secondary reflections, which would have a negative effect on hologram production.

Standard household mirrors have the reflective coatings on the back so that the glass protects the delicate reflective coating. Since front-surface mirrors don't have this protection, its important to handle them carefully. Fingerprints or any kind of dirt render the mirror ineffective. Handling the surface of the mirrors, even if your hands are clean, will leave oily fingerprints.

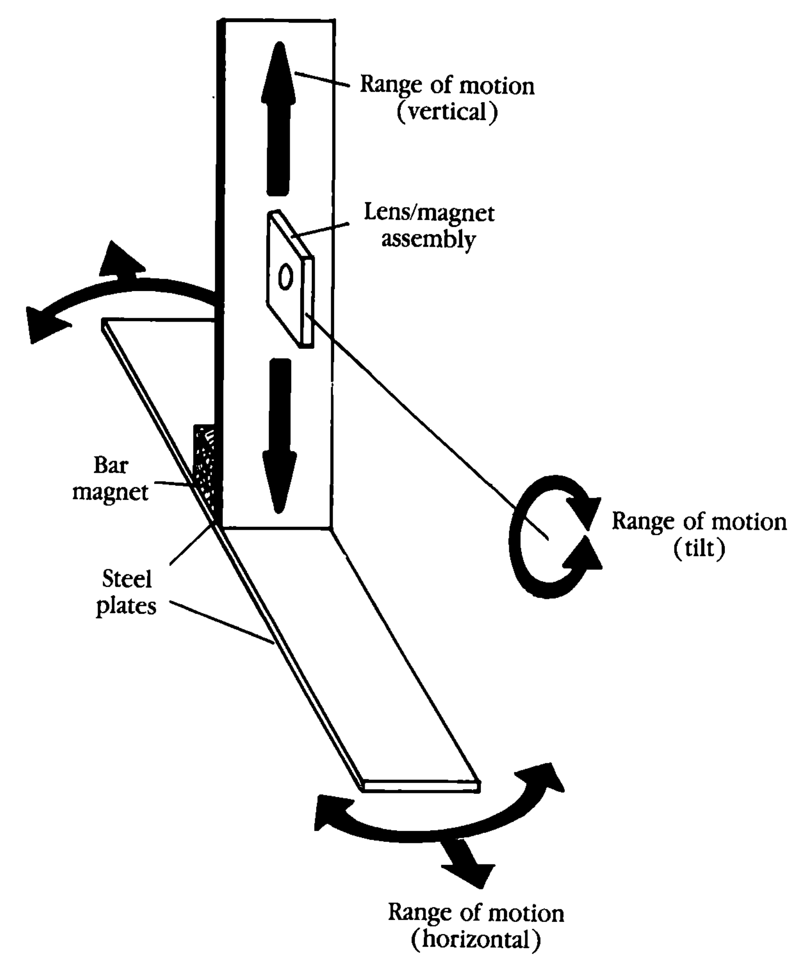

4-7 Mounting lens/magnet assembly on stand and adjustment axes.

To prevent fingerprints and oil from your skin from marring the surface of the mirror, handle the mirrors only from the edges. For handling larger mirrors, it is wise to purchase an inexpensive pair of white cotton gloves and wear them whenever handling them.

It might appear from the illustration that you could take a household mirror, turn it around, and have a front-surface mirror. Unfortunately, a black protective compound is usually painted on the back to protect the coating and to block light from back scattering.

Front-surface mirrors are available from Edmund Scientific, but an even better source for small front-surface mirrors is Meredith Instruments or MWK Co. Both sell small mirrors that can be glued directly to the magnets. They also sell long rectangular surplus mirrors that are about 1 inch wide by 9 inches long at ¼ inch thick. These mirrors are an excellent buy if you are capable of using a glass cutter; if so, you can make 8 or 9 useable transfer mirrors from one long mirror.

4-8 Magnet/lens assembly on stand.

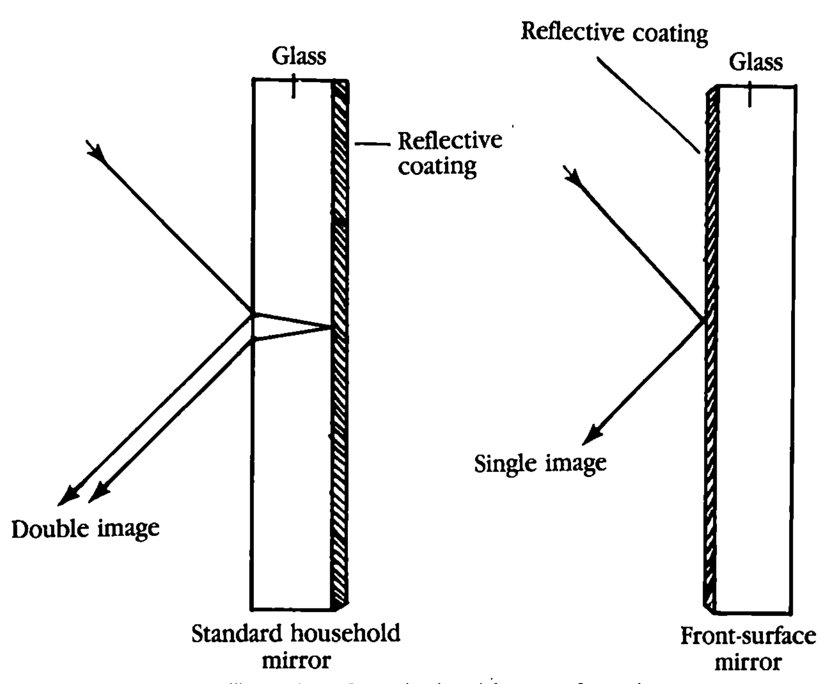

4-9 Illustration of standard and front-surface mirrors.

Cutting glass is easy, even if you have never done it before. I recommend learning to cut glass as it will save you money in the long run. Not only with front-surface mirrors, but with other optical components like beam splitters (see Appendix F "Cutting Glass").

Front-surface mirrors, whether cut to size or purchased, are glued to a small bar magnet as shown in FIG. 4-10. Complete lens adjustment through all axes is illustrated in FIGS. 4-11 and 4-12. This makes aligning transfer mirrors on the table a snap. Two front-surface mirrors on the T can be used to adjust the laser beam height (see FIGS 4-13 and FIG. 4-14).

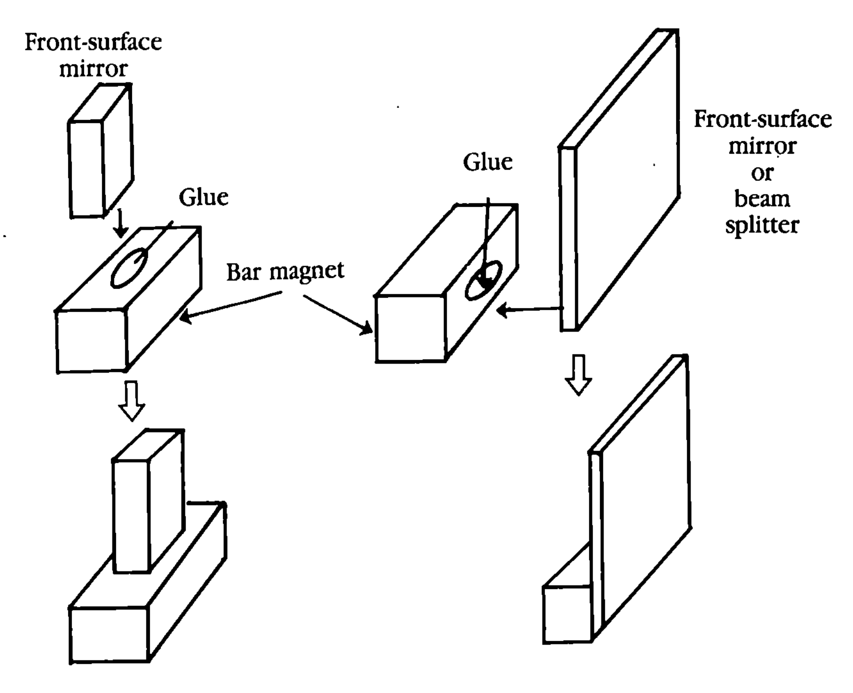

4-10 Mounting front-surface mirror or beam splitter.

Beam splitters

As its name implies, a beam splitter divides a incident beam of light into two well-defined beams. Part of the beam travels through, while the other is reflected (see FIG. 4-15). The beam splitter is a simple optical device consisting of a piece of glass with parallel faces. One glass face of the beam splitter has a partially reflective coating. This reflective material determines the ratio of reflective light to transmitted light. Beam splitters are made with a variety of such ratios. A 9:1 beam splitter transmits 90 percent of the incident light while reflecting 10 percent. A 1:1 beam splitter transmits 50 percent and reflects 50 percent. These are fixed-ratio beam splitters.

A variable beam splitter ratio changes across the surface. This allows holographers to tune the beam ratio to obtain optimum beam intensities. These types of beam splitters are expensive. For the purposes of this book, a few fix-ratio beam splitters are sufficient.

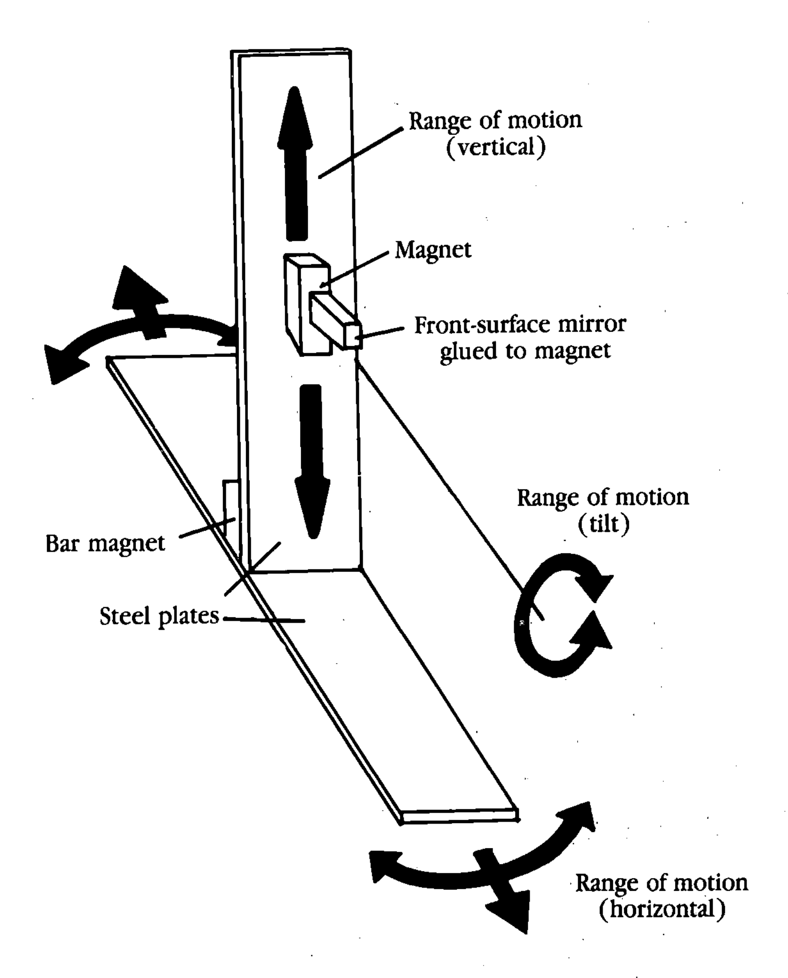

4-11 Adjustment axes with mounted mirror or beam splitter.

Inexpensive beam splitters are available from Edmund Scientific. Some of the beam splitters are made of glass the size and shape of microscope slides. These beam splitters can be cut with a glass cutter to make two or three beam splitters from one. Beam splitters are also glued to a rectanglular magnet in the same way as the front-surface mirror. With the beam splitter, the part that extends past the magnet is the working end.

Spherical mirrors

The spherical mirrors are glued to the rectangular magnet as in FIGS. 4-16 and 4-17. The concave surface of the mirror is facing away from the magnet. These spherical mirrors spread the laser beam exceptionally well. More information is given on setups in later chapters. There are other optical components to be mounted, but for the time being, this is what you need to get started.

4-12 Mounted mirror on stand.

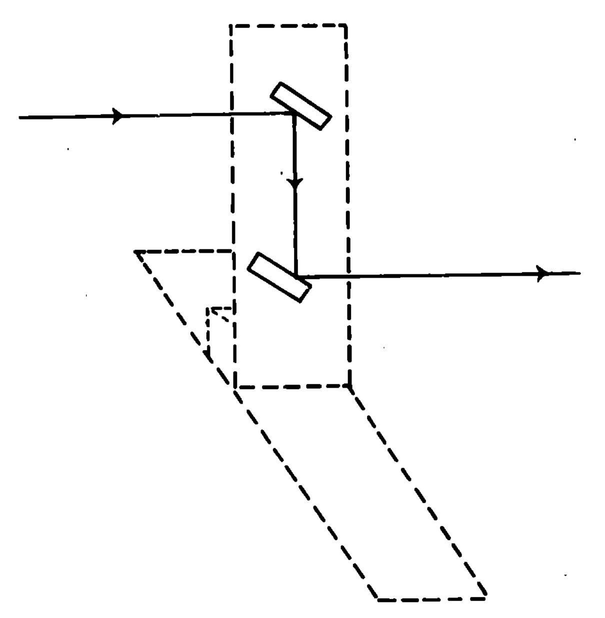

4-13 Mirror configuration to adjust beam height.

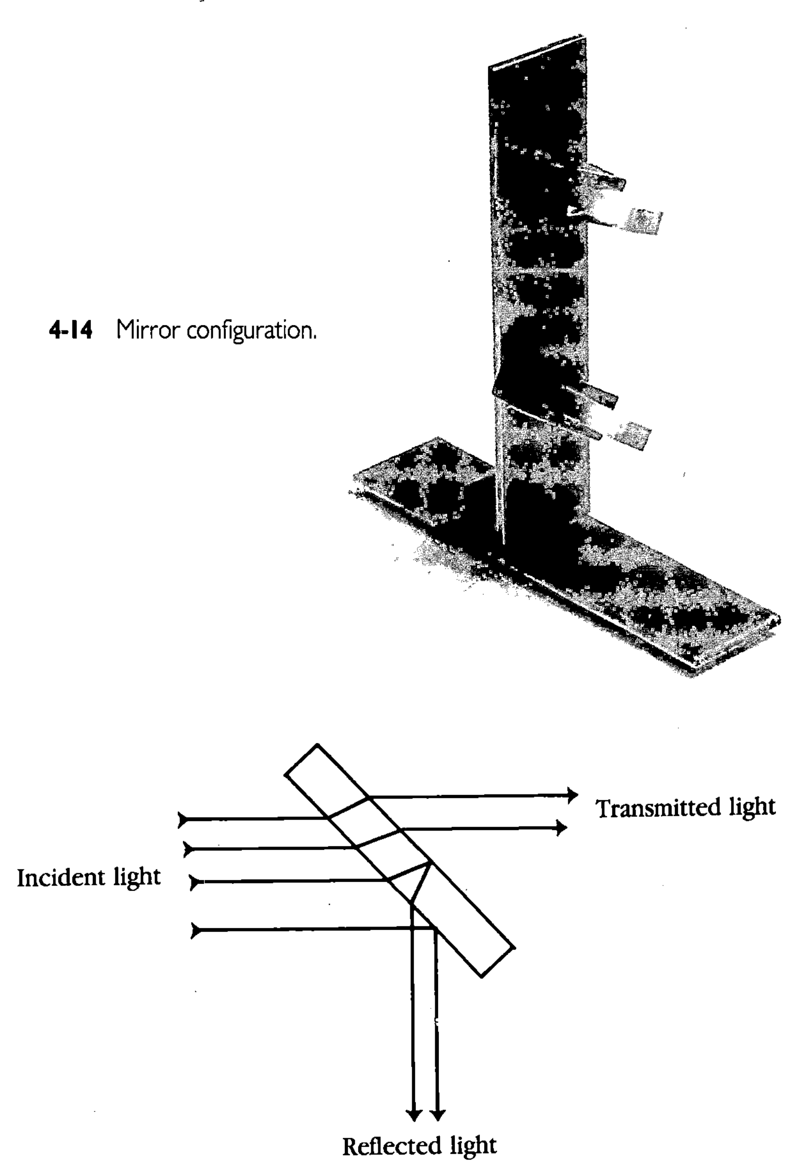

4-14 Mirror configuration.

4-15 Operation of beam splitter.

CLEANING OPTICAL COMPONENTS

Sooner or later, no matter how careful you are, you are bound to get a fingerprint or some dirt on an optical component. Whether it's a lens, beam splitter, or front-surface mirror, the procedure to clean is the same.

To remove dust, purchase a small can of compressed air. These are sold in photographic supply stores. They come in a variety of sizes, but purchase the smallest one available. To remove dust and debris, gently blow it off using the compressed air.

4-16 Operation of spherical mirror.

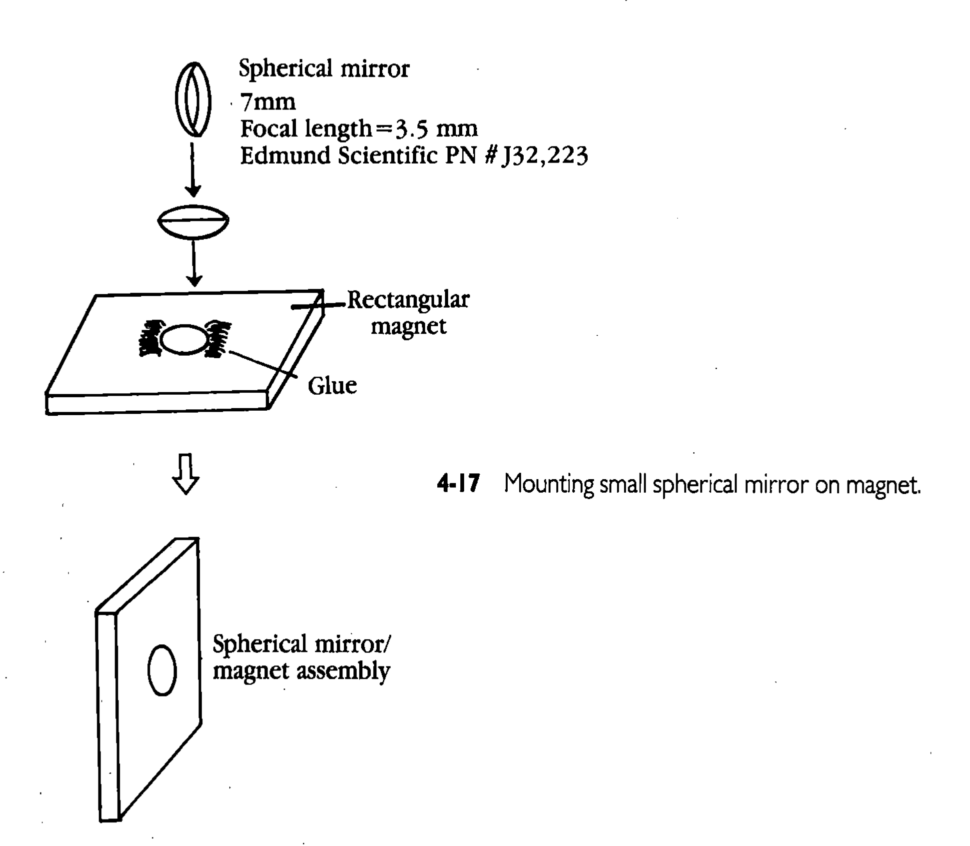

4-17 Mounting small spherical mirror on magnet.

To remove fingerprints and dirt, purchase cotton swabs, lens-cleaning paper, and lens-cleaning fluid or alcohol. Wrap one sheet of the lens-cleaning paper around the end of a cotton swab. Place a few drops of cleaning fluid or alcohol on the component. Using the paper-covered end of the cotton swab, gently wipe and clean the component using either a circular motion or a figure-eight motion. Allow the component to dry, and blow off any lint or dust with compressed air. The reason for using a lens-cleaning tissue paper is that it will not leave a lot of lint on the optical component after you're through cleaning off the dirt.

FILM PLATE HOLDER

With the use of rigid glass film plates, the film holder is very simple. The film holder consists of two office binder clips with a magnet glued to each one (see FIG. 4-18). The binder clips are available from any store that sells office supplies. When gluing the magnet to the binder clip, make sure the bottom of the magnet is parallel with the clamp to make it easier when securing a film plate. You don't *have* to glue the magnet to the clamp because the clamp is made of steel and the magnet will stick to it, but I find it easier to glue them together so I don't have to worry about looking for a magnet when I'm setting up.

4-18 Holographic film plate holder setup.

If you purchased the 12-inch-square steel plate for the table, you can eliminate the small steel bottom plate shown in FIG. 4-18. The clamps will stick directly to the 12-inch steel plate on the table. TABLE 4-1 shows the parts list for the film plate holder.

Table 4-1 Film Plate Holder Parts List

Quant.	*Item*	*Part No.*	*Supplier*
3	Spherical mirror	J 32,223	Edmund
1	Negative lens	J 96,134	Edmund
1	Positive lens	J 32,318	Edmund
1	Beam splitter (1:1)	J 31,411	Edmund
1	Beam splitter (3:7)	J 31,412	Edmund
1	Beam splitter (9:1)	J 31,416	Edmund
10	Rectangular magnets	mag-1	Images Co.
10	Steel plates	st-1	Images Co.
10	Bar magnets	mag-2	Images Co.
4	Front-surface mirrors		Meredith, MWK, or Edmunds

Chapter 5

Film

There are a number of different types of films available. I recommend you start with only one type of film but in two different sizes. The film is Agfa-Gevaert 8E75 HD NAH glass plate, which has an emulsion coating on one side of the glass. The HD suffix stands for *h*igh *d*efinition and NAH stands for *n*on*a*nti*h*alation. This film is suitable for reflection- as well as transmission-type holograms and gives excellent results. Finally, it is easier to use than the flexible acetate-backed (plastic) film because if you are using flexible acetate film, you must first make a film holder to keep the film rigidly in place (remember the problems with vibration). Although some film holders can be simple, for example two pieces of glass with a flexible hinge, the holder itself creates internal reflections caused by the glass (see the section on thin film interference in chapter 11, "Light"), which degrades the overall quality of the hologram. Experienced holographers use an index matching fluid between the film and the glass to alleviate the problem. However, to get you started with the minimum amount of fuss, use the glass plates.

The glass plates are a little more expensive than the acetate film. To reduce cost and waste, start with 2.5 × 2.5-inch film plates. These plates cost approximately $1.25 each. After you have some experience, you can start shooting the larger 4 × 5-inch plates. The cost of these plates is about $5 each.

When you first start shooting holograms, you have to anticipate ruining a couple of plates before you start producing good holograms. Using the smaller, less expensive film plates produces less anxiety about getting it right. Never berate yourself if a hologram doesn't come out because everything is an opportunity to learn. The idea is to have fun not give yourself an ulcer.

ANTIHALATION COATING

Although the plates used in this book aren't manufactured with an antihalation coating, other films are. It therefore becomes important to know what it does before you use other types of films. The emulsion coating on the film or plate is transparent. The antihalation is an opaque coating on the back of the film. It prevents light from back scattering into the emulsion after it has passed through the emulsion. It also prevents any light from entering the emulsion from the back side. This is advantageous when shooting transmission-type holograms because it cuts down reflections from the back of the plate. However, this film cannot be used to make white light reflection holograms. Recall from the geometric model that white light reflection holograms have beams interfering from both sides of the emulsion. Film with antihalation back have the suffix "AH," while film without it uses NAH. See TABLE 5-1 for a list of other films. The specifications of Agfa-Gevaert holographic films are atypical of film specifications in general. See TABLE 5-2 for the specifications of these films so you can use that information when checking the specifications of other films from other manufacturers.

Table 5-1 Films

Manufacturer	*Item*	*Sensitivity*	*Hologram*	*Film/plate*
Agfa-Gevaert	8E75 HD	Red	Both	Both
	8E56	Green	Both	Both
	10E75	Red	Transmission	Both
	10E56	Green	Transmission	Both
Ilford	SP673	Red/blue	Both	Both
	SP672	Green/blue	Both	Both
	SP737	Red	Reflection	Film

Eastman-Kodak also produces a line of film for holography, but these films are only available in England.

Table 5-2 Agfa-Gevaert Films

8E75 HD	Grain size is 35 nm. Resolution is 5000 lines/mm. Sensitive to red-light-emitting lasers.
10E75	Grain size is 90 nm. Resolution is 3000 lines/mm. Sensitive to red-light-emitting lasers.
8E56 HD	Same properties as 8E75 HD, but sensitive to green-light-emitting lasers.
10E56	Same properties as 10E75, but sensitive to green-light-emitting lasers.

Grain size and speed

Notice from TABLE 5-2 that the grain size is related to the resolution of the film. The grain size represents the size of the photosensitive silver particles in the emulsion. It might appear that because of the tiny size of the grains, the resolution of the film should be greater (14,000 lines/mm, with a 35 nm grain) than the quoted resolution. However, when developing the film, the grains of silver clump together, forming larger grains. This effectively reduces the ideal resolution to the practical resolution quoted.

EXPOSURE ENERGY AND RECIPROCITY FAILURE

Figure 5-1 is a graph that plots exposure energy to the developed plate density. The wavelength of light used to create the chart is 627 nm, which is slightly shorter than the HeNe laser of 632.8 nm. For all practical purposes, the chart represents HeNe light as well. The energy represents total energy exposure to the plate without regard to time. As an example, to achieve a density of about 1.5 for 8E75 film, you could make a one-second exposure of 20 $\mu J/cm^2$, or a two-second exposure at 10 $\mu J/cm^2$.

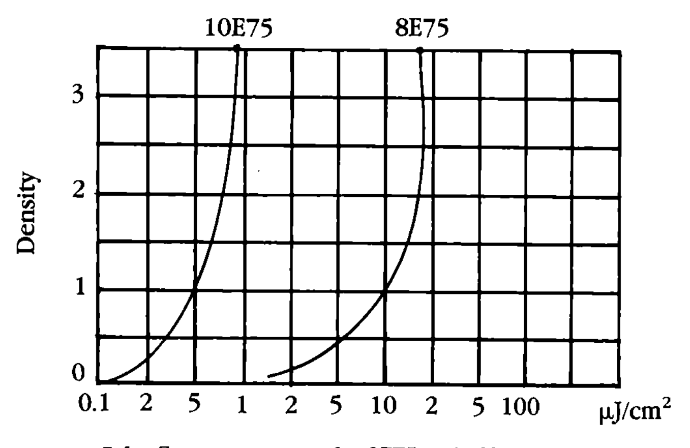

5-1 Exposure energy for 8E75 and 10E75 films.

Reciprocity failure is a characteristic of the emulsion in that it becomes increasingly less sensitive with very low light levels or very long exposures. Calculated exposure energies that would give proper densities have to be doubled or tripled with long exposures to compensate for the reciprocity failure. Fortunately, reciprocity failure is not a concern with this introductory volume. We are not making very long exposures, but it is a good idea for you to be at least aware of this film characteristic in the event that you start shooting holograms with long exposures.

NOTE: The Agfa specification sheet on the 8E75 film only mentions reciprocity failure for very short exposures, which is the norm with pulse ruby lasers. The information on reciprocity failure for long exposures is a general explanation for photographic films and is assumed to be similar for holographic films. More exact information isn't given, but exposures longer than 30 seconds probably have to be compensated.

DENSITY

When exposing film to produce a hologram, you should be aiming to develop a hologram to a particular density that will provide the brightest image. The *density* of a hologram represents the amount of metallic silver produced in the emulsion by exposure and development at the stage after developing but before bleaching. It is the degree of light-stopping ability (opacity) of the emulsion and is expressed as the logarithm of this opacity.

Figure 5-2 shows a filter that transmits 50 percent of the light incident upon it. To calculate the density of the material:

$$\frac{\text{Incident light (\%)}}{\text{Transmitted light (\%)}} = \frac{100}{50} = 2$$

$$\text{Density} = \log 2 = 0.3$$

The filter therefore has a density of 0.3. The first part of the calculation gives the reciprocal of the transmitted light. Take the base ten log of the answer to get the density.

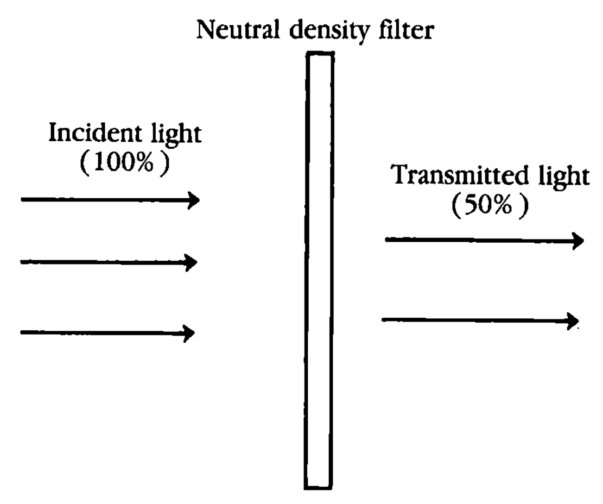

5-2 Operation of neutral density filter.

A filter that transmits 25 percent of the light incident upon it is:

$$\frac{\text{Incident light (\%)}}{\text{Transmitted light (\%)}} = \frac{100}{25} = 4$$

$$\text{Density} = \log 4 = 0.6$$

In holography, as a general rule of thumb, when developing transmission holograms try to achieve a density of 1.5 before bleaching. For reflection holograms, aim for a density of 2.5 before bleaching. For transmission holograms with a 1.5 density, approximately 3 percent of the incident light is transmitted.

$$\frac{\text{Incident light (\%)}}{\text{Transmitted light (\%)}} = \frac{1}{.03} = 33.3$$

$$\text{Density} = \log 33.3 = 1.5$$

For reflection holograms with a density of 2.5, approximately 0.3 percent of the incident light is transmitted.

$$\frac{\text{Incident light (\%)}}{\text{Transmitted light (\%)}} = \frac{1}{0.003} = 316$$

$$\text{Density} = \log 316 = 2.5$$

These descriptions on the percentage of light transmitted might be misleading, transmitting feeble amounts of light that are barely visible. However, this is not so because the eye perceives brightness in a logarithmic way. A 1.5 density appears (on the average) to transmit about 50 percent of the light incident upon it. Because of this illusion, neutral density filters are used to accurately gauge the density of the hologram before bleaching.

Neutral density filters

Because density is used in photography, I contacted a large photo supply house in New York City to purchase a "density chart." After being bounced around to three different departments, no one had ever heard of such a thing and only offered to sell me grey scales, but since they weren't calibrated I didn't think I could use them. After doing some research and talking to a few people, I found that I needed to use neutral density filters.

Neutral density filters are used in black-and-white and color photography to reduce the light intensity by a definite ratio without affecting the basic tonal values of the original scene. The filters are grey pieces of film calibrated in densities ranging from 0.1 to 4. Kodak makes a range of neutral density (ND) filters. TABLE 5-3 shows densities from 0.1 to 4.0 and the corresponding percent of light transmission. Add two ND filters together to obtain any desired density. For example, adding a 1.5 ND and a 0.9 ND gives a density of 2.4.

Table 5-3 Density vs. Light Transmission

Density	*Transmission (%)*	*Density*	*Transmission (%)*
0.1	80.0	0.8	16.0
0.2	63.0	0.9	13.0
0.3	50.0	1.0	10.0
0.4	40.0	2.0	1.0
0.5	32.0	3.0	0.1
0.6	25.0	4.0	0.01
0.7	20.0		

ND filters can be purchased locally or through mail-order companies such as Arkin-Medo or Spiratone (see Suppliers appendix). ND filters are a little expensive; a 3 × 3-inch (75 × 75 mm) plastic gel can cost $12 to $22 apiece. What I did to save money was to purchase one 0.5 ND filter that measured 3 × 3 inches. I cut it into nine 1-inch squares. See FIG. 5-3. By adding the filters together, I obtained higher densities. For the 1.5 density I laid three 1-inch square filters on top of one another and taped them to my safelight. For a 2.5 density I used 5 filters. When processing the film plate, I can hold the plate up

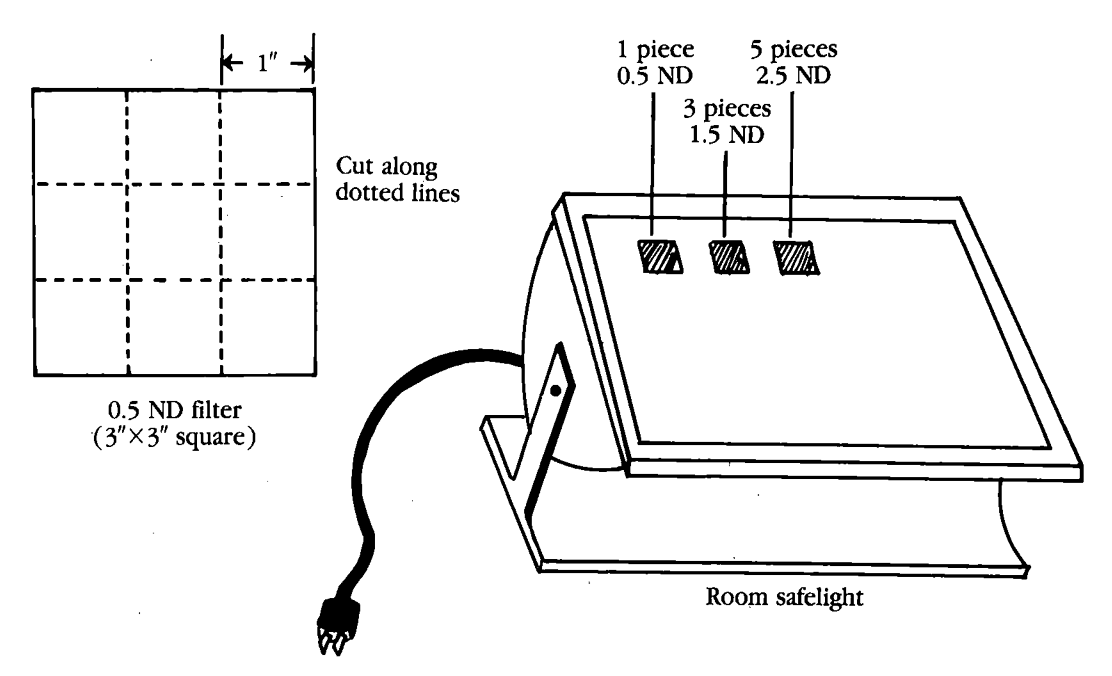

5-3 Cut pattern for ND filter and application to safelight.

to the safelight, next to the ND filter, and check its density. To be sure your safelight is in fact safe, see the section on Safelights later in this chapter.

The densities of 1.5 and 2.5 are important values in developing holograms. The 1.5 density is used when making transmission holograms. You try to expose and develop the transmission film plate to match the 1.5 density. The 2.5 density is used when developing reflection holograms. You then stop the development once the plate reaches the proper density and then rinse and bleach it.

ENERGY CONVERSION

The energy measurement in FIG. 5-1 is in μJ/cm^2. The μ stands for *micro* (10^{-6}), and J stands for joules (pronounced "jewels"). The joule is a measure of energy equal to 1 watt per second, which is the energy of 1 watt over the time period of one second. The cm^2 stands for centimeters squared. As an example, if a 5 mW beam is spread out with a lens to a diameter of 8 inches, the power per square centimeter (cm^2) is 15 μW. If the plate is illuminated at this distance for 1 second, the exposure would be 15 μJ/cm^2. *Erg* is also an energy measurement sometimes used to describe film characteristics. One erg equals 10^{-7} joules or 10 ergs = 1 μJ.

SPECTRAL SENSITIVITY

Figure 5-4 graphs the spectral sensitivity of the emulsion. Notice the sensitivity dips around 500 nm. That is the color green. This is the reason we can use a green safelight with the film. Also notice that the film becomes sensitive around

5-4 Spectral sensitivity of 8E75 film.

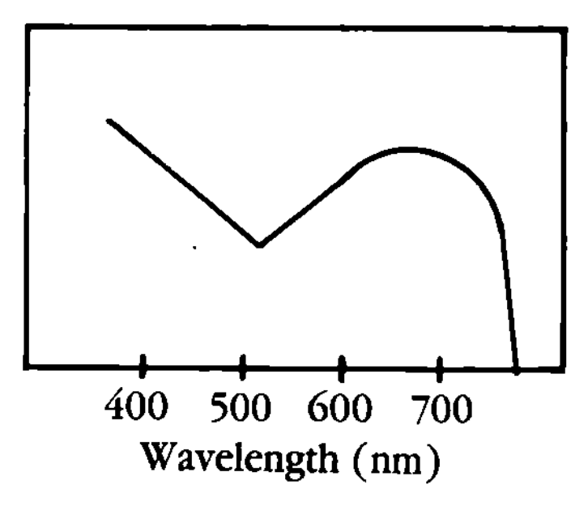

450 to 400 nm. This is the blue-to-violet region of the spectrum. So although Agfa quotes the film as being red sensitive, it is also blue sensitive and could also be used with the blue line from other types of lasers such as an argon-ion laser.

OTHER EMULSIONS

All the information given on film represents photosensitive silver halide emulsions. The following is a partial list of other emulsion types. The advantages of these other types include from being grainless, higher efficiency, erasable, or used to create stamping plates for mass-produced holograms.

Photopolymers Light-sensitive plastic polymer. Seen as a possible replacement to DCG emulsions.

Photoresist Used to make stamping plates for embossed holograms. Sensitive to shorter wavelengths.

Dichromate-gelatin (DCG) Easy to make, hard to work with, and sensitive to blue/green laser light. High light efficiency. More recently Jeff Blyth has made a DCG that is sensitive to red HeNe light and is easier to work with than the standard DCG that most are familiar with.

Thermoplastic Interference fringes recorded in the material are erasable.

SAFELIGHT

As shown in FIG. 5-4, the 8E75 holographic film is least sensitive around the 500 nm (green light) wavelength. This property of the film can be used to construct a safelight. A safelight provides sufficient illumination when working in a darkened room when setting up your table and/or during film processing. It allows you to see what you are doing without exposing or fogging the film.

You can make your own safelight from a flashlight by replacing the clear plastic front piece in the flashlight with green transparent plastic (see Suppliers appendix). You will probably need to use six or seven layers of green plastic to dim the light sufficiently for it to be safe. Unscrew the front piece of the flashlight. Remove the reflector and circular transparent plastic (front piece). You can use either the plastic or the outside diameter of the reflector as a pattern to cut the green plastic (see FIG. 5-5). When you're finished cutting several of the plastic gels, reassemble the flashlight with the green plastic gels in place of the original clear plastic.

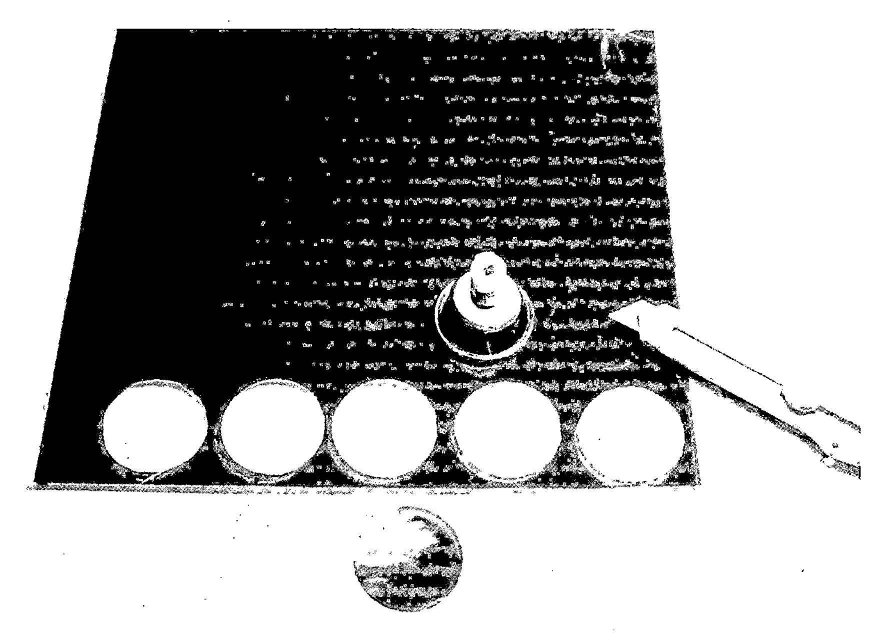

5-5 Cutting safelight material to fit flashlight.

Because of the amount of filtering, the light will not be very visible in an average lighted room. Let a minute or so pass when in a darkened room to allow your eyes to adjust before the illumination given off by your safelight appears bright enough to work by.

Many photography supply stores sell inexpensive safelights for use in a darkroom. To make it useable in your holography work, replace the standard red filter in the photographic safelight to green. Figure 5-6 shows a converted

5-6 Modified room safelight and safety flashlight.

standard flashlight and a converted darkroom safelight with green filters for use in holography. If you can't find an inexpensive safelight locally, try mail-order houses like Spiriatone (listed in the Suppliers appendix). To find other excellent sources for photographic supplies, I advise purchasing a copy of *Shutterbug* magazine. This magazine is published monthly and is predominately advertisements for mail-order houses

Working under a safelight isn't as bad as it first might seem. Most of the setup work can be done under regular lighting. After the setup is complete and you're ready to place the photographic plate on the table (last thing you do), you turn off the room light and turn on the safelight.

Checking the safelight

In complete darkness and with the safelight off (if you open the film box to remove a film plate with the safelight on and it's not "safe," you might fog all your film) remove a plate, close the box, and then turn on the safelight.

The emulsion is only on one side of the glass plate, and the plate should go into the developer with the emulsion side up to ensure good coverage with the developer solution. So, to check for which is the emulsion side, slightly moisten a finger and press it against the glass plate. The sticky side is the emulsion side. Keep in mind that any residue left on the emulsion will leave a smudge. So use only the smallest section possible from a corner of the plate when checking it.

Having identified the emulsion side, put on rubber gloves and pour the developer solutions into a tray (see chapter 6, "Developing holograms" for more information on solutions and amounts). Place the plate, emulsion side up, into the tray. Shine your safelight on the plate from a distance of about 6 inches. The longer the plate remains clear, the safer the light is. If the plate turns black, that means the film has been exposed. You should be able to keep the light on the plate a minute or two (or longer) without fogging the plate. Naturally, the time elapse before the plate begins to fog is the safest amount of time the plate can be exposed to the safelight. When using the safelight to place the film plate on the table, turn it off or point it in another direction after the film is in place.

If you are using a room safelight and are using the same green filter material that's in the flashlight, the room safelight is probably safer because it is usually a greater distance from the film and will have less impact. If in doubt, check this safelight also. Use the same procedure as described previously.

Don't let the same dog bite you

When I first started making holograms, I encountered a problem. The holograms I was producing were dim, partial, or nonexistent. I tried to localize the problem. At first, I thought the table wasn't stable enough. I checked and rechecked it with the interferometer. It seemed perfectly stable. Then I checked the lens mounting, the film plate mounting, and the model mounting. Then I painted the model silver, thinking it wasn't reflective enough. Nothing I did had any positive effect on the image quality. During this time, I looked at the safelight once or twice, thought I should check it, then quickly decided not to. After all, I had three green gels, which must be plenty. After I exhausted every possibility, I was left with nothing to check but the safelight.

I dumped a fresh plate into the developer and shined the safelight onto it. Within 3 seconds the plate fogged out. I doubled the number of gels in the flashlight to six pieces and started shooting holograms (ahem, without rechecking the safelight). The image quality improved, but it was still rather dim. I decided to add some more green gels just to be sure. But instead of getting a deeper green, the color shifted to a very dim red. I checked the safelight again, and sure enough it fogged the film within 10 seconds. I remembered I had some green transparent material from another manufacturer. To the eye, the color and density of this other green material was identical to the first, but I decided to change it anyway. After placing seven gels in the flashlight, I noticed the filtered light remained green; it didn't shift to red as with the other material. I considered this a good sign and tested the material. After a one-minute exposure in the developer, the film didn't fog. I moved the flashlight closer, about an inch away from the film. Another minute passed, and the film still didn't fog. I was starting to get worried. Did I put water into the developing tray instead of the developer? I let another minute pass, and there was still no fog visible on the plate. To check to see if I really put developer in the tray, I turned on the room light. The plate blackened almost instantly. Thereafter, my holograms' image quality improved dramatically. Now I have a "safe" safelight. The moral of the story is to check your safelight.

Chapter 6

Developing holograms

The following information on film processing details the chemistry and procedures for developing transmission and reflection holograms. Although plates are used here, the instructions and chemistry are the same for acetate films.

CHEMISTRY

You have two options: you can purchase a development kit that contains all the chemicals you need to develop holograms, or you can purchase the four basic chemicals and do it yourself. The development kit is available from the Photographers Formulary (see Suppliers appendix). Other than the four basic chemicals, it includes preservatives that increase the shelf life of the solutions. I advise "budding" holographers to purchase the kit (see FIG. 6-1). The kit has the chemicals in premeasured packets. No measuring is required; all you do is add the chemicals to water. Plus, the shelf life of the solutions is quite long (a few months or more). The image quality produced from the kit is no better than the image quality produced from using the four basic chemicals, but the best reason to buy the kit is that it's easy to use because you don't have to measure any chemicals. The Photographers Formulary sells three different holographic kits; get the JD-2 kit, which is suitable for reflection and transmission holograms (as of 3/90, the JD-2 kit costs $12).

If you prefer to buy the four basic chemicals yourself, the individual chemicals initially cost more than the kit, but in the long run, it is much cheaper to process holograms mixing your own.

Getting started

Whether you purchase the kit or buy the chemicals, follow these simple guidelines when mixing the chemicals.

- Use only distilled water to make your developing and bleach solutions because it has the least amount of impurities. Any impurities in the water can have a negative impact on the developed image quality of the hologram. This is especially true for city dwellers who have a municipal water supply. Distilled water is not spring, mountain, naturally carbonated, or sodium-free water. If your local supermarket doesn't carry it, try a local auto supply store (distilled water is used in car batteries and radiators).

- Don't use metal containers or utensils. Use only glass or plastic utensils and containers. Some of the chemicals are corrosive to metal and would therefore react with it, rendering the solution worthless.
- Wear rubber gloves when mixing the chemicals and when developing holograms. If chemicals make you uncomfortable, also wear a rubber apron and goggles.
- Be neat. Clean all spills at once. Rinse off all utensils once you are finished with them. Label all solutions, and mark the date the chemicals were mixed.

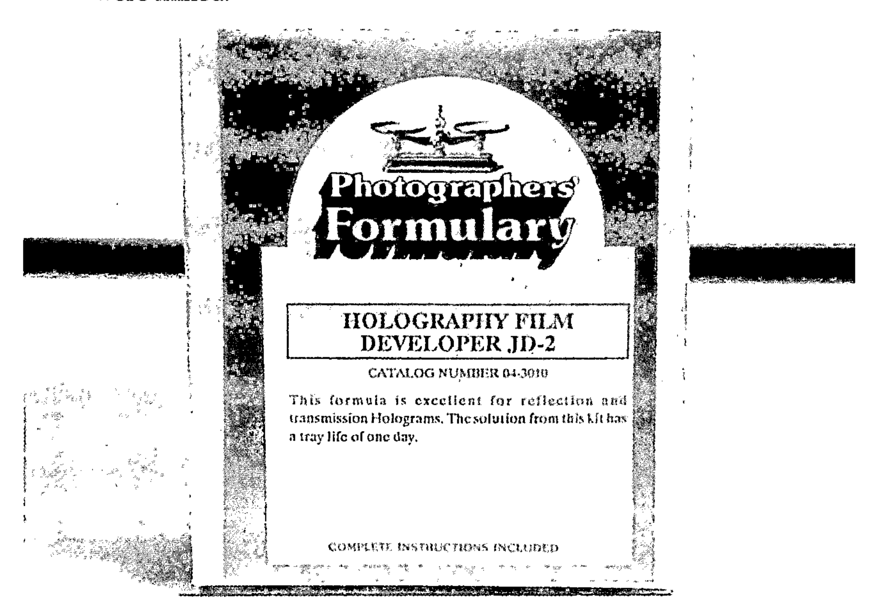

6-1 Hologram development kit available from Photograhers Formulary.

To begin, you need plastic containers to store the solutions in. Photo supply stores sell plastic bottles just for this purpose. Buy a minimum of three bottles, each with a 1-liter capacity. If you have purchased the holographic development kit, mix the chemicals according to the directions supplied with the kit and store them in the plastic bottles. Again, try to use distilled water. The kit is designed to make a liter of three stock solutions. A liter of water is equal to about 4½ cups. Use a measuring cup that is designed for use in cooking.

The following instructions are for those of you who plan to purchase and mix the chemicals yourselves. If you have purchased the development kit, you can skip over these instructions and go on to the next section ("Gloves").

Mixing your own

The four chemicals you need to develop transmission and reflection holograms are (see FIG. 6-2):

- Pyrogallic acid
- Sodium carbonate
- Potassium dichromate
- Sulfuric acid (concentrated)

6-2 The four basic chemicals you need to develop holograms.

These chemicals are available from Photographers' Formulary, Arbor Scientific Co., and Images Co. (see Suppliers appendix). In addition to the storage bottles, you'll need one plastic (or glass) measuring cup and one set of plastic measuring spoons, the same type of cup and spoons you would use for cooking. The formulas are given in grams of chemical per liter of water, but since I don't expect anyone to have a scale that measures grams, conversions to cups and teaspoons are provided. Using measuring spoons isn't as accurate as weighing the chemicals, but it is close enough.

All measurements are based on level, not heaping, spoons of chemicals. When measuring dry chemicals with a spoon, the actual weight will vary, depending on how well packed the chemical powder is, but the formulas allow for these deviations.

Bleach First mix the bleach. Since it has a long shelf life, mix a liter of solution:

1 liter (4½ cups) distilled water
4 milliliters (½ tsp.) concentrated sulfuric acid
4 grams (½ tsp.) potassium dichromate

When adding the acid to the water, be careful. Add the acid to the water over a sink in case it spills, measuring ½ cup of water in your plastic 1-cup measuring cup. Using the plastic measuring spoon, measure ½ level teaspoon of acid and carefully add it to the water. Rinse and dry the teaspoon. Take ½ level teaspoon of potassium dichromate and add to the water. Pour the solution into the storage bottle. Add 4 cups of additional distilled water to the storage bottle to bring the solution to 1 liter.

Developer The developer is called a "pyro" developer (the process is called *pyrochrome*). Store the developer as two parts, labeled A and B. When you are ready to use the developer, mix equal parts of A and B in the development tray just before developing the hologram.

Pyro developer A:

1 liter (4½ cups) distilled water
10 grams pyrogallic acid

The pyro developer doesn't have a long shelf life, perhaps about 24 hours before it oxidizes and becomes useless. Therefore, mix the pyro developer just prior to shooting holograms. If you are doing a lot of shooting, prepare a batch of pyro early to use throughout the day. But more likely, you'll be shooting only a few holograms at a time, so because of the short storage life and the fact that we are starting with small holographic plates (2.5 × 2.5-inch), it would be wasteful to mix a liter at a time. Therefore, only mix about 100 milliliters of each part (A and B). Using the development trays for 2.5-inch-square plates, 100 milliliters is sufficient to do two processing runs. Each run can process two holograms.

The volume of 100 milliliters is equal to ½ cup of water. To ½ cup of distilled water add ¼ teaspoon of pyrogallic acid. Store in a plastic bottle.

Pyro Developer B:

1 liter (4½ cups) distilled water
60 grams sodium carbonate

Although this part of the pyro developer doesn't go bad as quickly as the other part, it should still be made at the same time (not stored). To make 100 milliliters of part B, Add 1 level teaspoon of sodium carbonate to ½ cup of distilled water. Store in a plastic bottle. As you proceed in holography, you might need larger stock solutions, so since there are 1000 milliliters to a liter, simply multiply the 100 ml out to whatever volume you need.

Gloves

You must wear gloves when developing holograms because the chemicals are toxic and can be absorbed through the skin. You will need to put your hands into the solution to lift the holographic plate from one tray to another. I use rubber household gloves that are sold in most supermarkets.

Developing trays

In terms of the amount of chemicals required for developing, the small 2.5 × 2.5-inch plates are best accommodated by a 3.5 × 5-inch tray. You can use transparent plastic picture frames sold at many hardware and photo stores. Look for ones with sides about an inch high. Purchase a minimum of three trays. Remove the white cardboard filler and discard it. Scratch the inside surface of the tray with coarse sandpaper and scrape some deep gouges in the inside with a screwdriver or some other tool to help prevent the plate from sticking to the bottom of the tray when it is filled with developing solution.

It's a good idea to practice moving a film plate in and out of a tray that has some liquid in it. My opinion is it's worth ruining one holographic plate to practice. Remove one plate from the box (remember to open the box and take a plate out in absolute darkness until you have tested your safelight and determined it to be really safe). Take the plate (don't forget to close the box) and check for the emulsion side (see chapter 5). After you're confident you know which is the emulsion side of the plate, fill the trays with ½ cup (100 milliliters) of water each. Put on your rubber gloves because you'll be wearing them when you do the actual developing. Practice putting the plate in the tray, rocking the tray back and forth slightly to agitate the film, then removing the plate and placing it in the next tray. When you're comfortable with the process, you're ready to develop your first hologram. NOTE: When you start processing larger (4 × 5-inch) holograms, you can purchase the same type of clear plastic tray in a 5 × 7-inch size.

DEVELOPING YOUR HOLOGRAM

Before detailing the procedure to develop holograms, it is important to check your safelight. Put 50 milliliters each of part A and part B development solutions in a tray. Remove a single plate (in absolute darkness) and place it emulsion side up in the tray. Remember to put on your gloves before touching any solutions. Shine the safelight on the plate and make a note of the time it takes for the plate to blacken or fog. (See chapter 5 for further instructions on checking the safelight.)

Developing

Arrange three trays in front of you (see FIG. 6-3). The tray on the right is for the pyro developer solution. The center tray has distilled water. The tray on the left has 100 milliliters of bleach. Remember to wear your gloves before putting your hands in any solutions.

Under a safelight, mix 50 ml (¼ cup) each of your stock pyro A and pyro B solutions in the first tray. The mixed pyro solution has a useful lifetime of about 15 minutes, so *you should be ready with the exposed plate before you mix the two solutions in the tray.* The plate is developed for 2 minutes in the developer tray. Place the exposed plate, emulsion side up, into the developer. Gently rock the tray back and forth to keep fresh solution in contact with the plate. Proceed for about 2 minutes. If you have the neutral density filters, you can compare the plate density to the filters.

Remove the plate from the developer and place in the distilled water for 30 seconds. This step isn't mandatory, but it will extend the life of the bleach so you can reuse it. After the time has elapsed in the rinse, place the plate in the bleach tray. Rock it gently back and forth as before. Keep the plate in the bleach

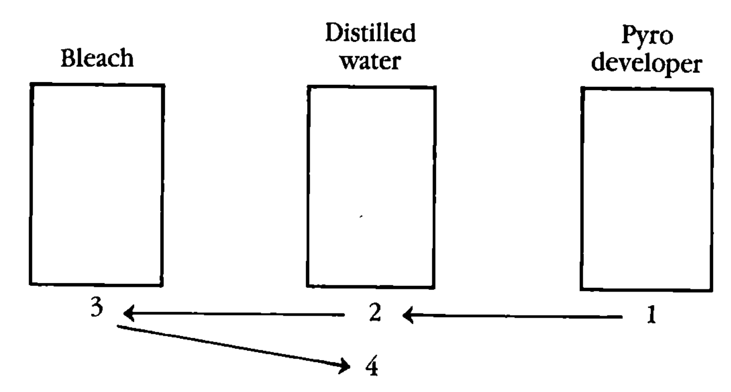

6-3 Tray arrangement to develop holograms. Under safelight illumination, follow these steps, as numbered in the illustration: 1) In pyro developer for 2 minutes; 2) Rinse in distilled water for 30 seconds; 3) In bleach until plate clears (about 1 minute); 4) In rinse tray, place in sink and bathe in running water for 5 minutes.

until it becomes completely clear again, which usually takes about 1 minute. After the plate clears, the emulsion is light safe, so you can turn on the room light.

Remove the plate from the bleach and put it back into the rinse tray. In the sink, run tap water until it reaches about room temperature and place the tray under the running water for 5 minutes. After the final rinse, remove the plate, stand it vertically against a wall, and allow it to dry. The holographic image will not be visible until the hologram is completely dry.

Plates sometimes dry with water spots on them. You can dip the plate in Kodak Photo-flo solution (mix according to directions) after the final rinse to prevent water spots from forming. Some people wipe the plate with a photographic squeegee to remove excess water and thereby speed up drying. Others use a hair dryer to quicken the drying time. If you use a hair dryer, set it on warm or low, or you could damage the hologram with excessive heat.

The bleach usually stays good for quite a few processings. Even so, it isn't a good idea to mix the used bleach with fresh bleach. Instead, when you're finished developing your holograms, keep the used bleach in a small plastic container and continue to reuse it until it is spent; bleach is spent when it takes over a minute to clear the hologram.

The tray size is 3.5 × 5 inches, so it is large enough to process two plates at once. Do not lay one plate on top of the other; place them in the tray so that they both lie on the bottom of the tray with the emulsion side up. This way, you can process both plates at once. You do not have to add any additional chemistry to process two plates. The amount you measured for a single plate are sufficient to process both plates.

I usually work alone, so it can become a problem to keep track of how long I have a plate in the developer. Photo stores sell large clocks with luminescence hands, but these are rather expensive. My solution was to build a simple audible timer. The timer beeps every 5 seconds. This allows me to agitate the film without having to try to look at my watch to check the elapse time. Looking at your watch in a darkroom with gloves on can be a real hassle. The plans for the audible timer is in Appendix B.

TROUBLESHOOTING

If you have followed the directions for mixing the chemicals properly, the chemistry should not present any problems. This section covers problems with exposure time of the plate when shooting the hologram. It is assumed you have checked your safelight and it is indeed "safe."

If the plate turned black very quickly (within 15 seconds), the plate is overexposed. You need to either cut your exposure time when the plate is on the table or spread the beam wider to reduce the laser light intensity. If the plate doesn't darken or acquire any density during development, it is underexposed. Increase your exposure time on the table.

If during developing, the exposure is off, you do have some latitude when developing the plates. For instance, if the plate turns black quickly, it doesn't serve any purpose to keep the plate in the developer solution for the full two minutes. Remove the plate from the developer and continue to rinse and bleach; you could still salvage a good hologram. Likewise, if the plate is very light and hasn't developed to the proper density, try keeping the plate in the developer for a longer time.

DISCARDING USED CHEMICALS

The developing chemicals are poisonous. Check with your local government (sanitation department) on how you should handle getting rid of the spent developer and bleach (the rinse water is diluted enough to empty down the drain). Don't just empty the spent chemicals down the drain unless they say it's okay to do so. If the chemicals are discarded in the drain, dilute them with a lot of running water. Sanitation departments often have a specific location that handles this type of waste, or they might tell you to just put the containers out with the regular trash.

If you do have to bring it to a particular location, it isn't practical to go every time you decide to develop a hologram. Accumulate the spent chemistry in plastic gallon containers. When you have accumulated a few containers, make a trip and get rid of all of them. Keep the discarded bleach and developer in their own container; don't mix them.

Most of the time, since you will only have small amounts of waste, the sanitation department won't bother to charge you. Or if they do, it will only be a nominal charge, like $2 for 10 gallons of waste.

TECHNOLOGICAL ASPECT OF DEVELOPMENT

This section contains a basic description of the process involved in developing a hologram. It isn't required reading, but it's here if you want to know more details about the chemistry of holograms.

Overview

In development, the pyrogallic acid develops the latent image in the holographic emulsion by reducing the exposed silver halide to metallic silver. The metallic silver is dark and provides the density of the hologram.

The bleach is called a *reversal* bleach. Instead of removing the undeveloped silver halide as in conventional photography, the bleach dissolves and washes away the metallic silver, hence the name "reversal."

You would think that by washing away the metallic silver you would lose the image, and in conventional photography you would be correct. Recall that in a hologram, it is the interference pattern that reconstructs the image. The undeveloped silver halide that is left in the emulsion is a copy (a positive) of the same interference pattern as the metallic silver. The former, clear interference pattern is able to reconstruct a brighter image than that of metallic silver because it absorbs less light.

Absorption and phase holograms

If you stopped development after placing in pyro developer, you would have created an *absorption hologram.* The dark metallic silver interference patterns reconstructing the image would have absorbed a good deal of the light. Because of this, it is called an absorption hologram.

When the hologram is bleached, it transforms into a *phase hologram.* It's called a phase hologram because once the metal silver was removed, the interference pattern is made of clear undeveloped silver halide emulsion. Essentially, the fringes are caused by the difference in the refraction index of the undeveloped silver halide emulsion and the emulsion material that has been washed off the metal silver. The phase hologram reconstructs a brighter image because the dark metallic silver fringes that absorb so much light have been removed by the bleach.

Basic photochemistry

An *emulsion* is a homogenized mixture of gelatin and silver halides. *Homogenized* means that the tiny grains of the insoluble silver halides are suspended equally throughout the gelatin. The emulsion is usually coated on a plastic (acetate) or glass base. The gelatin is called a *binder.* It supports the silver halide crystals in suspension yet permits the diffusion of water, chemicals, and waste products in and out of its structure. It is this characteristic of gelatin that enables various photochemical reactions to take place such as developing and bleaching.

Silver halides Silver halides are light-sensitive salts. In general, a *salt* is defined as a compound composed of metal and nonmetal elements. Silver can form with halides such as chlorine, bromine, and iodine. These compounds are called silver halides or silver salts. Sodium can also form with the halide group, for example sodium chloride (table salt). It is evident from the appearance of table salt that salt forms into crystals. The silver halide crystals are very small. The size of the crystal is related to the film speed. The smaller the crystal, the slower the film. While the sodium halides are soluble in water, silver halides are not.

The silver salts are photosensitive. To explain, when silver combines with a halide, there is a release of energy. If Ag is the chemical symbol for silver and Br is the symbol for bromine, the following relationship holds:

$$Ag + Br \rightarrow AgBr + \text{energy}$$

This is fairly simple chemical reaction that shows silver combining with bromine to form silver bromide plus a release of energy. What is interesting is that the process is reversible. If light strikes the silver halide, the process is:

$$AgBr + \text{light energy} \rightarrow Ag + Br$$

Each silver grain in the emulsion consists of billions of molecules of silver bromide. When light strikes the grains, several bonds are broken, leaving free atoms of silver and bromide within the grain matrix. These grains correspond to the latent image, and in the case of holography, the fringe pattern. When the emulsion is placed into a developer, those grains that make up the latent image have a greater tendency to develop than those grains that are unexposed.

$$AgBr + \text{developer} \rightarrow Ag + Br + \text{oxidized developer}$$

In reality, silver bromide (AgBr) is the prevalent silver halide in holographic emulsions. Other silver halides are also present in the emulsion to a lesser degree.

Salts As stated, salts form when a metal atom combines with a nonmetal. Silver salts are not soluble in water, but many of the other salts used in holography are. When a typical salt dissolves in water, it disassociates into its metallic and nonmetallic parts. The metallic part separates carrying a positive charge, while the nonmetallic carries the corresponding negative charge. Table salt dissolving in water is represented as:

$$NaCl + \text{water} = Na^{+} + Cl^{-}$$

Acids and bases Acids disassociate, leaving a hydrogen ion H^{+}. An acid such as hydrochloride breaks down as:

$$HCl \rightarrow H^{+} + Cl^{-}$$

Bases disassociate with an ion called a *hydroxyl* (OH^{-}).

pH pH is the measurement of the acidic or basic nature of any solution. The scale runs from 0 to 14. A pH7 is midpoint and is considered neutral. Any solution with a pH less than 7 is an acid, while any solution with a pH above 7 is a base. A buffer is a compound designed to keep a solution at a particular pH level.

Oxidation and reduction Oxidation and reduction usually occur in tandem. As one material oxidizes, another reduces. A material *oxidizes* when it loses electrons and becomes a positive ion. A material *reduces* when it absorbs electrons. An emulsion developing is an oxidation-reduction reaction. The silver halide reduces to metallic silver, and the developing agent oxidizes.

In the pyrochrome process used to develop holograms, the pyrogallic acid reduces the exposed silver halide grains that form the latent image into metallic silver. The sodium carbonate is added into the developer solution because as the carbonate disassociates in water, it supplies OH^- (hydroxyl) ions. This neutralizes the H^+ ions produced by the pyrogallic acid-reducing the silver halide.

Solution of silver halides Silver salts are not soluble in water, but certain chemicals, such as sodium thiosulfate, dissolve silver halides by forming soluble complexes with the silver. In standard photographic development, this washes away the undeveloped silver, leaving the image (metal silver) in the emulsion.

The pyrochrome process of developing holograms leaves the silver halide and removes the metallic silver, a two-step process. The potassium dichromate in the bleach oxidizes the metallic silver into a compound that is soluble in the sulfuric acid. The sulfuric acid in the bleach washes away the metallic silver interference pattern. The silver halide that was not developed is unaffected by the bleaching process and remains in the emulsion. Although material is removed from the emulsion when washing the silver away, the emulsion material (gelatin) remains.

Pyrogallic acid $C_6H_6(OH)_3$ Pyrogallic acid is a poisonous chemical. It can be absorbed through the skin and can be fatal if large areas of the skin are exposed to it. Handle it with caution. Pyrogallic acid was the first practical developing agent and was used as early as 1851. This material still remains important in photography as well as holography.

Chapter 7

Single-beam holography

The preceding chapters have prepared you to shoot holograms. Let's begin now with a single-beam holography setup on the isolation table. It is a good idea to keep records concerning all of your setups along with the results to help you gain experience and learn from your mistakes faster. Your records show you what works and what doesn't work, and more importantly, what you need to do to get it to work. Whether you need to change an exposure time, optical component, or beam ratio, your records will be there for you to refer to when planning new setups. Figure 7-1 is an example of such a record that you can photocopy for your setups.

SINGLE-BEAM REFLECTION HOLOGRAM

Start with 2.5 × 2.5-inch plates. The exposure times given are calculated for this plate size. These times should put you in the ballpark, but you will probably have to adjust the exposure time a little to produce the brightest images for your environment. The easiest way is to check the density of the hologram while developing. Or if you are familiar with the procedure for making test strips, commonly used in photography for determining exposures, this method can be adapted to holography by masking and uncovering the plate for different exposure times.

A single-beam white light reflection hologram is viewable with white light and is quite impressive. Figure 7-2 illustrates the setup on the table. If you have the 12 × 12-inch metal plate (recommended in chapter 2) on the end of the table, it will be much easier to set up various configurations on the table because you can eliminate the bottom horizontal metal plate of the holographic film plate holder. The magnetic holders secure to the 12-inch-square metal plate itself. When using optical components on this end of the table, you only need the vertical metal plate of the inverted T. Place it alongside a magnet secured to the 12-inch-square metal plate.

Holographic Information Record			
Hologram ID / Name			
Date		Temperature	
Film Size & Type		Exposure Time	
Developer		Development Time	
Bleach		Bleach Time	
Laser			
Beam Strength Ratio	Ref	Obj.1	Obj.2
Sketch			
Results:			

Put Additional Information On Reverse Side

7-1 Holographic information record.

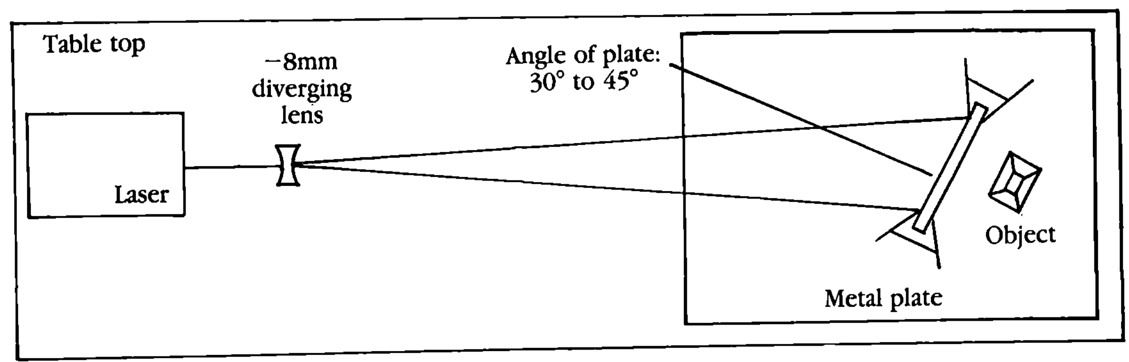

7-2 Setup for single-beam white light reflection hologram.

Depth of field

Single-beam reflection holograms have a restriction on the depth of field behind the plate that can realistically be holographed. This restriction is approximately one-half the coherence length of the laser. For the low-power (0.5 mW to 5.0 mW) HeNe lasers we are using, the coherence length is usually around 6 to 10 inches, which means you have a depth of field of 3 to 5 inches behind the plate. Figure 7-3 illustrates the depth of field that you can holograph in a single-beam setup.

The coherence length of the laser is a measurement of the distance the light will remain coherent with itself when it is split. To see this more clearly, examine FIG. 7-3B. The laser light is incident on the plate and passes through the emulsion (recall the emulsion material is transparent). The reason the depth of field is one-half the coherence length is that the light must travel twice the distance of the depth: first through the plate, to the object, then reflected back to the plate to meet the incoming light. So the round-trip distance the light travels is actually two times the depth.

The interference pattern is produced by the reflected light off of the object and the incident light on the plate. You should place whatever object you want

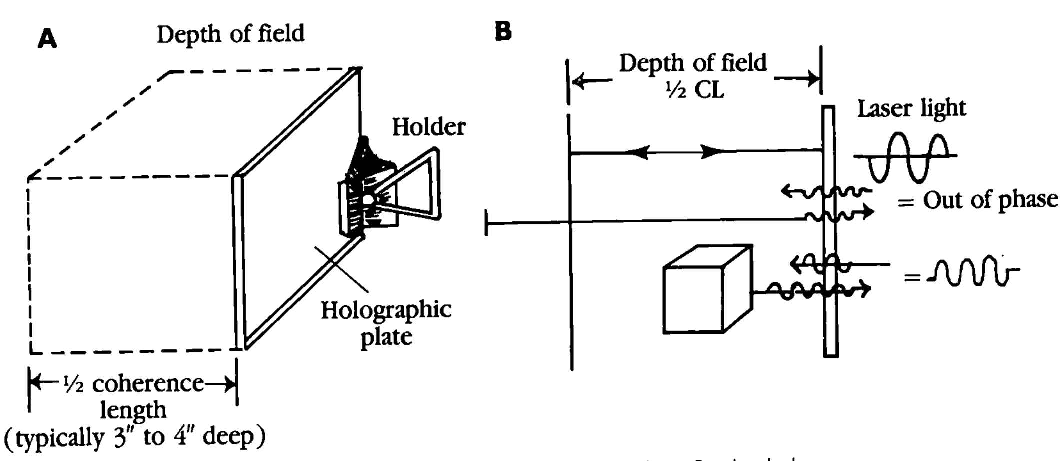

7-3 Depth of field in single-beam white light reflection hologram.

to holograph within this depth of field, preferably as close to the plate as possible to produce the brightest hologram. (Split-beam setups have a much greater depth of field, but that is covered later.) More information on the coherence length is in chapters 2 and 12.

Shutter card

You need something to block the laser beam as necessary to make the exposure. Typically, beginners use a shutter card. The shutter card can be made of any stiff, opaque material. Attach a binder clip to each side of the card at the bottom (see FIG. 7-4) to act as a stand for the card to hold it upright.

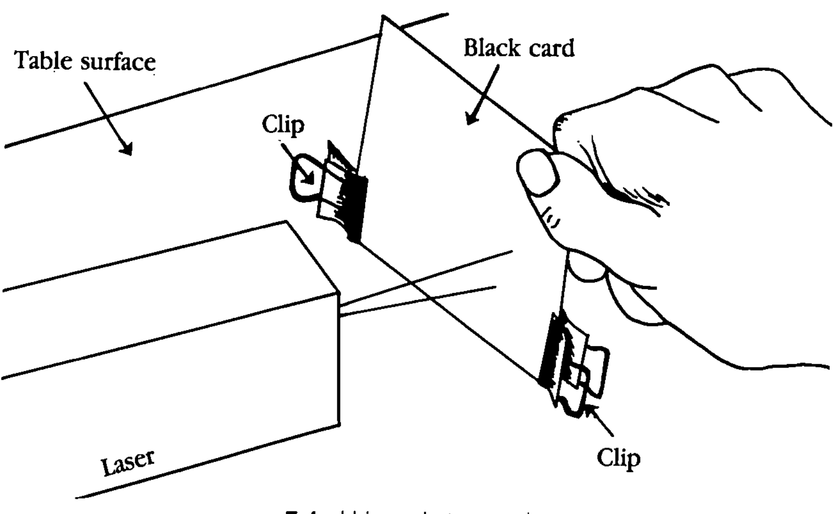

7-4 Using a shutter card.

Procedure

Choose an object you want to holograph. The object should be smaller than the plate and preferably a light color, white, off white, silver, or or other metallic finish. You can choose a darker item to shoot later after you have some experience, but to start with, holograph something that will show up brightly. The object should also be rigid, something that won't flex, bend, or move during the exposure. For my first object I chose a small white sea shell.

The object should be secured to prevent it from moving or rocking during the exposure. One of the easiest ways to accomplish this is by putting a small ball of clay on the table where you're placing the object. Push the object into the clay. Instead of clay, you could also use "Fun-Tak." Fun-tak is a blue-colored adhesive material with the consistency of clay. It can be pulled apart and pushed together again, rolled into balls, whatever. The material is reuseable, and because of its strong adhesive property is worthwhile to purchase. You can purchase it locally in hardware or dime stores.

Set the laser up on one end of the table opposite the metal plate. Turn on the laser. The laser should be warmed up (turned on) 30 minutes or more before you start shooting any holograms. This warm-up period is necessary for the laser to stabilize.

Position the −8mm lens close to the laser aperture. Hold a white card at the opposite end of the table to make the laser beam visible. Position and adjust the lens so that you have a good beam spread on the card.

If you have a used film plate, for example the one you practiced moving in and out of the development trays, position the plate in the spread laser beam so that it has the most even illumination possible. Placing a white card behind the plate can help determine the illumination. Secure the magnetic plate holders in that position to hold the holographic plate. Position and secure the object you want to holograph behind the plate. When you view the object through the plate, that is what your hologram will look like. Make any adjustments to the object to holograph it in the best position possible. Remember to keep the object as close to the plate as possible.

Remove the used plate, keeping the plate holders in position. Position the shutter card in front of the laser to block the beam. Turn on the safelight and shut off all other lights. Wait a minute to allow your eyes to adjust to the lower light level. Remove a fresh holographic plate from the box. Check for the emulsion side. Place the plate in the plate holders with the emulsion side facing the object.

Lift the shutter card off of the table, but not high enough to let the beam pass. Keep the card in this position for the relaxation time (about 30 seconds). To make the exposure, raise the card above the beam for the exposure time (about 1 second; see following list), and return the card to the table. The plate is ready to be developed (see chapter 6).

The exposure time for reflection and transmission holograms using 2.5-inch-square holographic plates with Agfa 8E75:

5.0 mW laser—1 to 1.5 seconds
2.5 mW laser—2 to 3 seconds
1.0 mW laser—5 to 7 seconds

For 4×5-inch holographic plates with a 6-inch-diameter beam spread, multiply the exposure time by 4.

SINGLE-BEAM TRANSMISSION HOLOGRAM

Figure 7-5 illustrates a single-beam transmission holographic setup. Notice how the beam geometry has changed from the reflection setup. With the reflection hologram, the interference pattern was generated from the reflected light off of the object (object beam) behind the plate and the incident light on the front of the plate (reference beam). Essentially, each beam struck the holographic plate from opposite sides.

With the transmission setup, both beams, object and reference, strike the plate from the same side. The geometry of this simple setup isn't ideal, but it does work. The illustration shows the shadow cast by the object to show that the object should be positioned so that its shadow doesn't fall on the plate. The procedure to expose the plate is the same as in the reflection hologram (described earlier in this chapter).

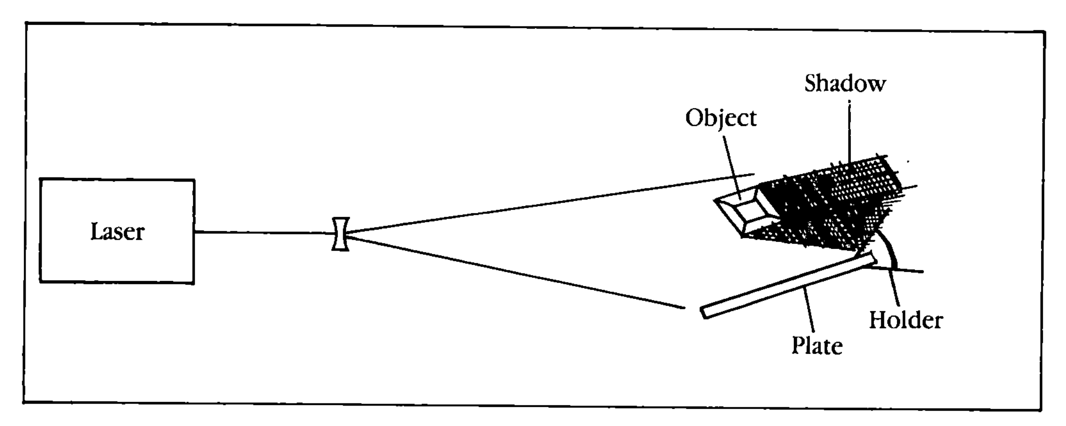

7-5 Setup for single-beam transmission hologram.

IMPROVED SINGLE-BEAM TRANSMISSION

The addition of a front-surface mirror as depicted in FIG. 7-6 improves the quality of the transmission hologram. The front-surface mirror scopes some of the incident light and reflects it toward the film plate. This reflected beam acts as the reference beam.

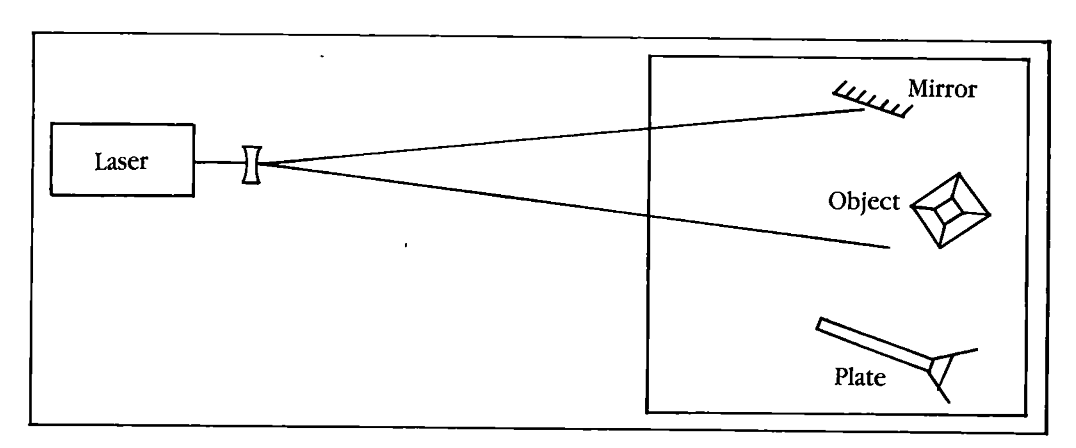

7-6 Setup for improved single-beam transission hologram.

Overhead mirror

I have designed a front-surface mirror holder, illustrated in FIG. 7-7. It can also be used for overhead reference beams in either reflection or transmission holograms. I am calling it an "overhead mirror" because it will likely be used more often in that capacity. See TABLE 7-1 for the parts list. Wear the white cotton gloves when handling the front-surface mirror to prevent oil and dirt from your hands from marring the reflective surface of the mirror.

Refer to FIG. 7-7. Drill a ⅜-inch hole in the two pieces of stock steel as shown. Glue the plastic bushing into the hole. Secure the glass hinges onto the front-surface mirror. Make sure that the pivotal rod of the glass hinge is exactly

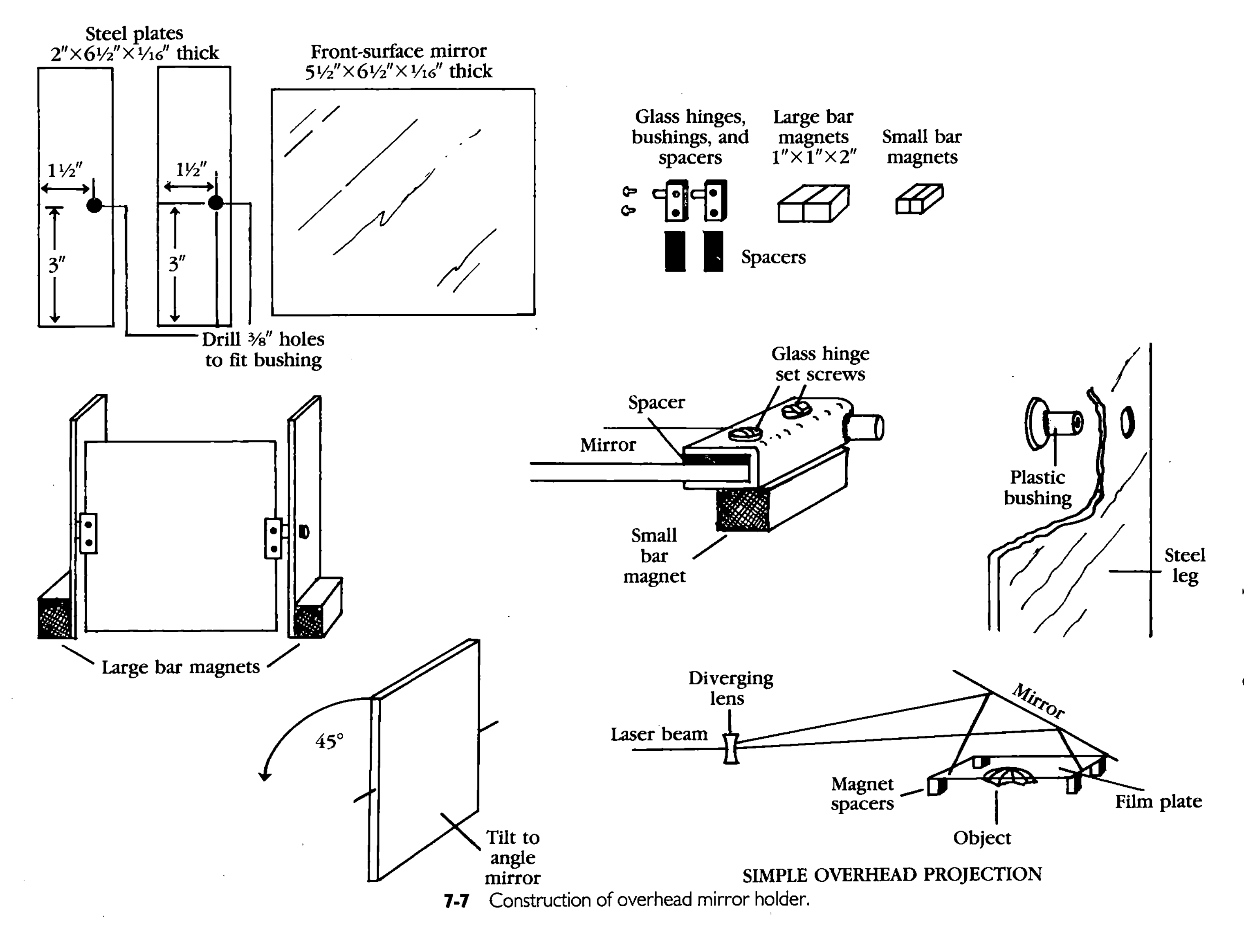

7-7 Construction of overhead mirror holder.

Table 7-1 Parts List for Overhead Mirror

Quant.	*Item*	*Supplier*
1	Front-surface mirror, 4″ × 5″ or larger	Edmund Scientific
2	Glass pivot hinges	Images Co.
2	Steel stock 2″ × 7″ × 1/16″	Images Co.
2	Small bar magnets	Images Co.
2	Large bar magnets	Images Co.
1	Pair of white cotton gloves	local store

at the midpoint on the mirror. Glue the two small bar magnets on top of the glass hinges, making sure that the magnets lie flush against the steel legs when the mirror hinge is inserted into the bushing so the magnets will hold the mirror at whatever angle you set it.

Having completed the construction, here is how to assemble the overhead mirror. Place one of the large bar magnets on the table where you want to set up the mirror. Place one leg against the magnet with the bushing on top. Place the mirror in the bushing. Bring the other leg over and place it so that the mirror's opposite hinges pass through the bushing. Bring the second large bar magnet over and secure it to base of the second leg. Check out the assembly. Make sure both legs are straight. The weight of the large bar magnets is sufficient to secure the assembly without a base plate. If the mirror is placed on the 12-inch-square metal plate, it will be that much more secure. If you have doubts about the stability of the assembly, they will probably disappear when you have to wrestle the assembly apart to store it.

You can use taller legs to raise the height of the mirror. Since the legs are the most inexpensive part of the assembly, you might want to make a couple of sets with various heights.

ALTERNATE SINGLE BEAM REFLECTION

The overhead mirror can be used as illustrated in FIG. 7-8. The advantage of this configuration is that the objects can be laid flat and mounted very securely. It's an excellent configuration for holographing jewelry and small objects. I holographed a seascape by filling a small rectangular tin with white sand and placing an assortment of small shells, coral, and ancient coins in the sand, making it look like the bottom of the ocean. It worked out well.

Position the film plate above the object with small wood blocks or magnets. Put a bit of clay or Fun-Tak on the edges of the blocks or magnets and place the plate on them so only the corners of the plate rest on the blocks. The clay prevents the plate from vibrating during exposure. If you use wood blocks, secure them in position with hot glue to prevent them from moving or vibrating. If you use magnets, they will be naturally secured to the metal plate. The magnets save a lot of time when making these setups, which is the main reason I recommended in chapter 2 that you get the 12-inch-square metal plate for the end of the table.

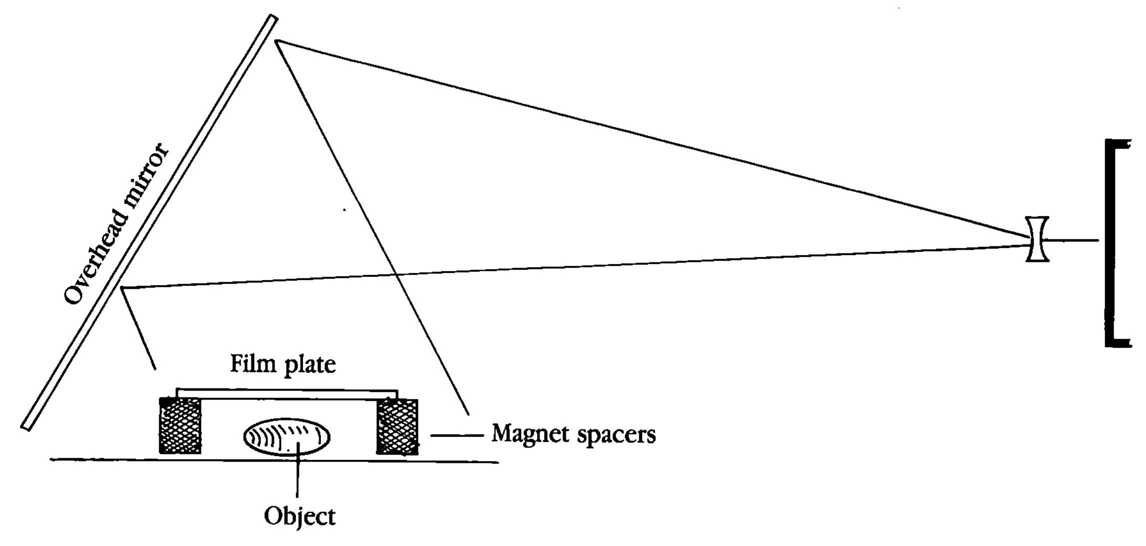

7-8 Alternate setup for single-beam reflection hologram.

Procedure

Adjust the diverging lens to spread the beam evenly on the mirror. Adjust the angle of the mirror to reflect the beam onto the plate assembly. It's a good idea to position a white card in place of the plate when adjusting the beam. The white card will allow you to see the beam's spread much more easily. When your mirror is set and the object is in place (don't forget to place a small amount of clay on the edges of the magnets), follow the procedure outlined previously for making an exposure. Since the beam is typically spread out a little further in this setup, the exposure time might be increased slightly.

IMPROVING THE QUALITY OF SINGLE-BEAM HOLOGRAMS

There are limitations as to what can be done to improve single-beam holograms. The most important aspect is the quality of the spread laser beam. You might have noticed when you tested your laser that the spread beam from the lens has whorls and marks. This is caused by dirt and imperfections on or in the lens. Some of them can be removed by cleaning the lens. The most common approach to improving beam quality is to use a *spatial filter*. A second approach, one I use extensively, is a *spherical mirror*.

Spatial filter

The spatial filter is a remarkably simple optical device in principle. Figure 7-9 illustrates a spatial filter. The lens of the filter focuses the laser beam to a point that coincides with the placement of a pinhole. Any light scattered by dirt or imperfections in the lens doesn't pass through the pinhole. The resulting beam from the spatial filter in quite clean and of optimum quality. Although the spatial filter is simple in principle, putting it into practice is another matter. The difficulty lies in aligning the lens so that its focal point is directly on the pinhole.

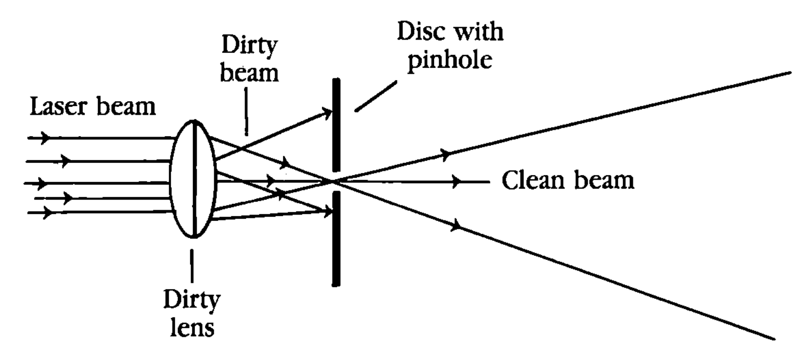

7-9 Operation of spatial filter.

The pinhole diameter is about 25 μm (0.001 inch), so if you are off by the slightest amount, the beam will be blocked by the filter. Commercial spatial filters use a microscope objective for the lens and have micrometer adjustments in all three axes. Appendix B explains how to build a simple spatial filter. The applications of the homemade spatial filter are limited, however, due to its simplistic construction. But it can be used for all single-beam setups using the 2.5-inch plates.

Spherical mirror

The small, spherical mirrors that were mounted on the rectangular magnets offered tremendous improvement over lenses. First, because of its short focal length, the beam spreads so quickly that you could shoot 8 × 10-inch holographic plates on the small table. Also, the quality and clarity of the spread beam is remarkable and is surpassed only by a spatial filter. Hence, I recommended buying three of these mirrors. They are also used extensively in split-beam work (chapter 9). Figure 7-10 illustrates a simple setup using the spherical mirror. Because the beam spreads so quickly, the transfer mirror and spherical mirror are located on the far end of the table with the plate. When you first align the mirrors, the spread beam might appear dirty. Simply move the spherical mirror until you find a clean spot on it. You can use the spherical mirror in all the single-beam setups shown so far. You might wonder why I didn't start using the spherical mirror in the first place. The reason is that it's more difficult to align the mirrors properly. I felt after you have succeeded in making a hologram with the lens, you would have more patience in aligning the mirrors and more practical experience in what you're looking for in a spread beam. The added bonus is that you would see the remarkable improvement in the quality of the hologram when using the mirrors.

Overhead beam The spherical mirror can also be used to project an overhead reference beam similar to the large overhead mirror (see FIG 7-11). Or, it can be used in conjunction with the overhead mirror to produce a greater beam spread for larger film plates. Why bother with the overhead mirror if the spherical mirror can accomplish the same thing? The answer is that they accomplish the same thing, but in limited applications. There are some setups where the overhead mirror will be the mirror of choice. In addition, knowing

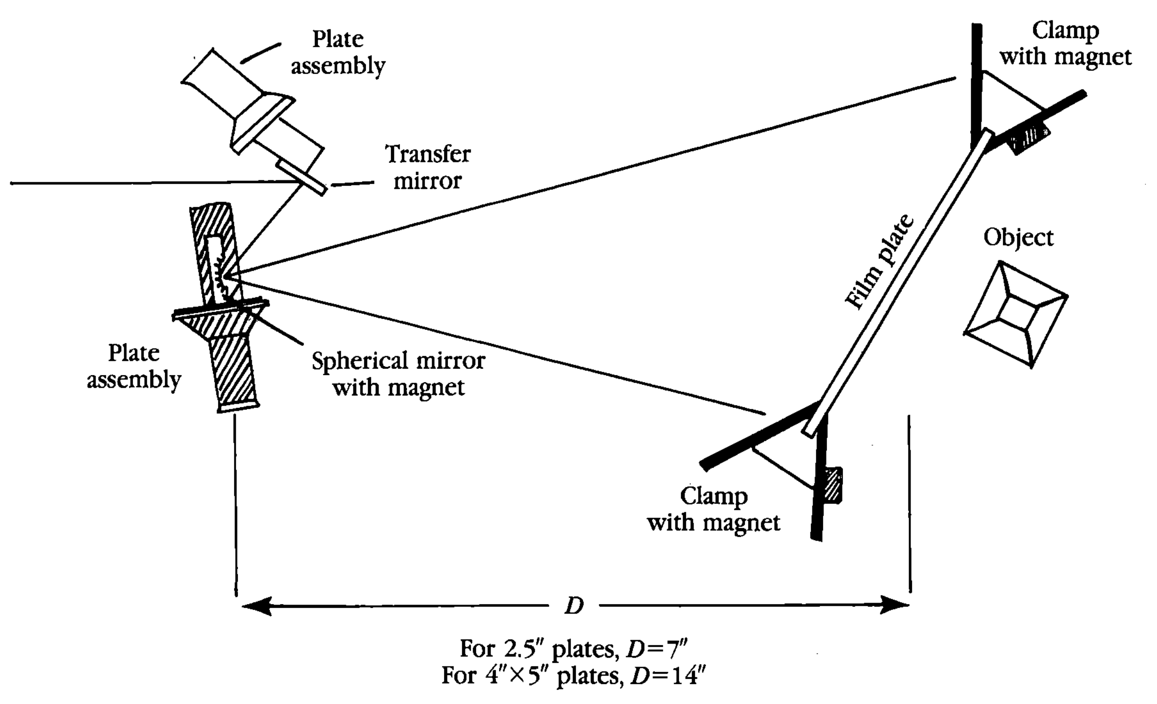

7-10 Overhead view using spherical mirror in place of diverging lens.

how to accomplish a specific task a number of ways creates flexibility, versatility, and innovation in your holography setups.

VIEWING HOLOGRAMS

After you have exposed and processed your holographic plate, it's time to take a look at your hologram. First let the plate dry because if it's still wet, you probably won't see anything. Each type of hologram requires its own type of illumination.

White light reflection

White light reflection holograms, as the name implies, are viewable in white light. The best type of illumination for this hologram is a point light source. The sun is a good example of such a source.

Tungsten halogen lamps are also an excellent light source. They have a good color temperature and a 2000- to 3000-hour lifetime. Images Co. sells tungsten halogen lamps specifically for illumination of reflection holograms. The lamps have a built-in glass reflector that focuses the light in front of the lamp. An inexpensive adapter is available to screw the lamp into a standard incandescent lamp socket. The cost of the lamp and the adapter is $25. Replacement lamps are available for $15.

Incandescent lamps can be used, but the image quality isn't as good as with the tungsten halogen. When using an incandescent lamp, notice that if you

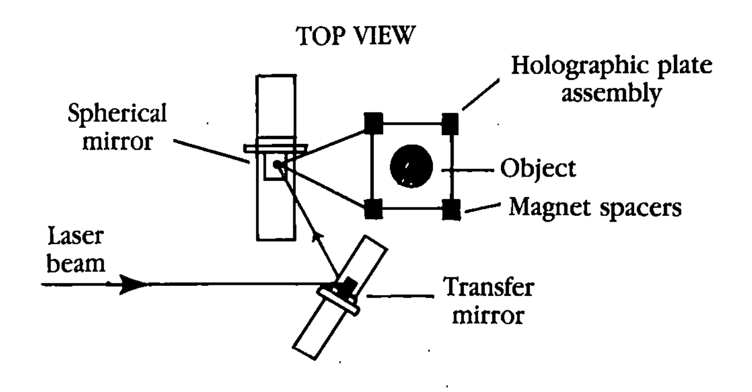

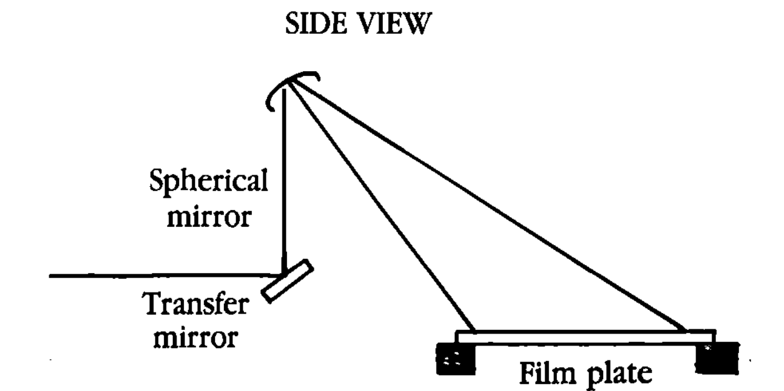

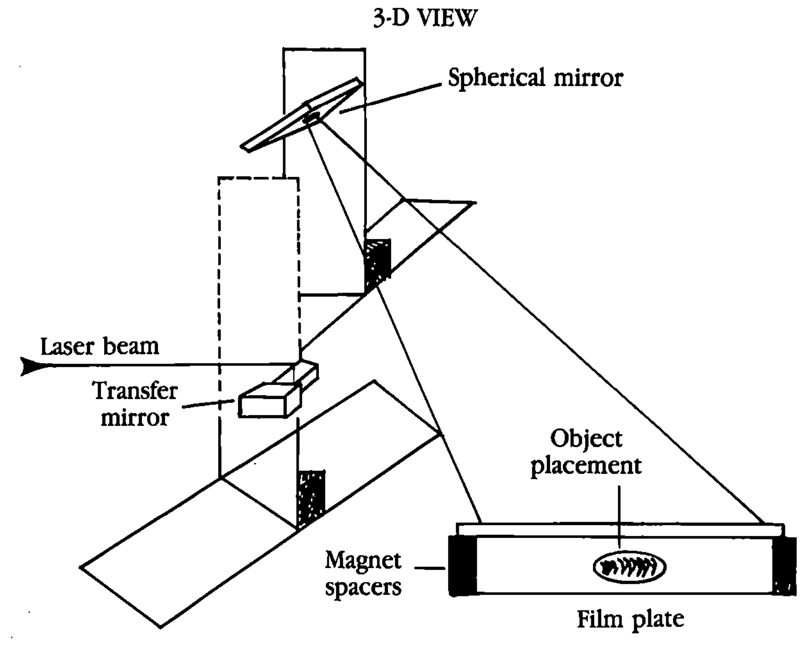

7-11 Using spherical mirror for overhead projection.

increase the distance of the hologram from the bulb, the image appears sharper. This is because the lamp becomes more like a point light source as the distance increases.

Reflection holograms work something like a mirror, so view the hologram from the same side as the light (see FIG. 7-12A). To see the image, move the plate around in the light until you find the best angle to view the hologram. Flip the hologram over to see the difference in the real and virtual images.

To improve the quality of the playback image, put a black sheet of paper behind the hologram. To make this effect permanent, spray paint the back of the hologram black. Do this only with reflection holograms you want to display because once it's painted, you can't use it to make copy holograms.

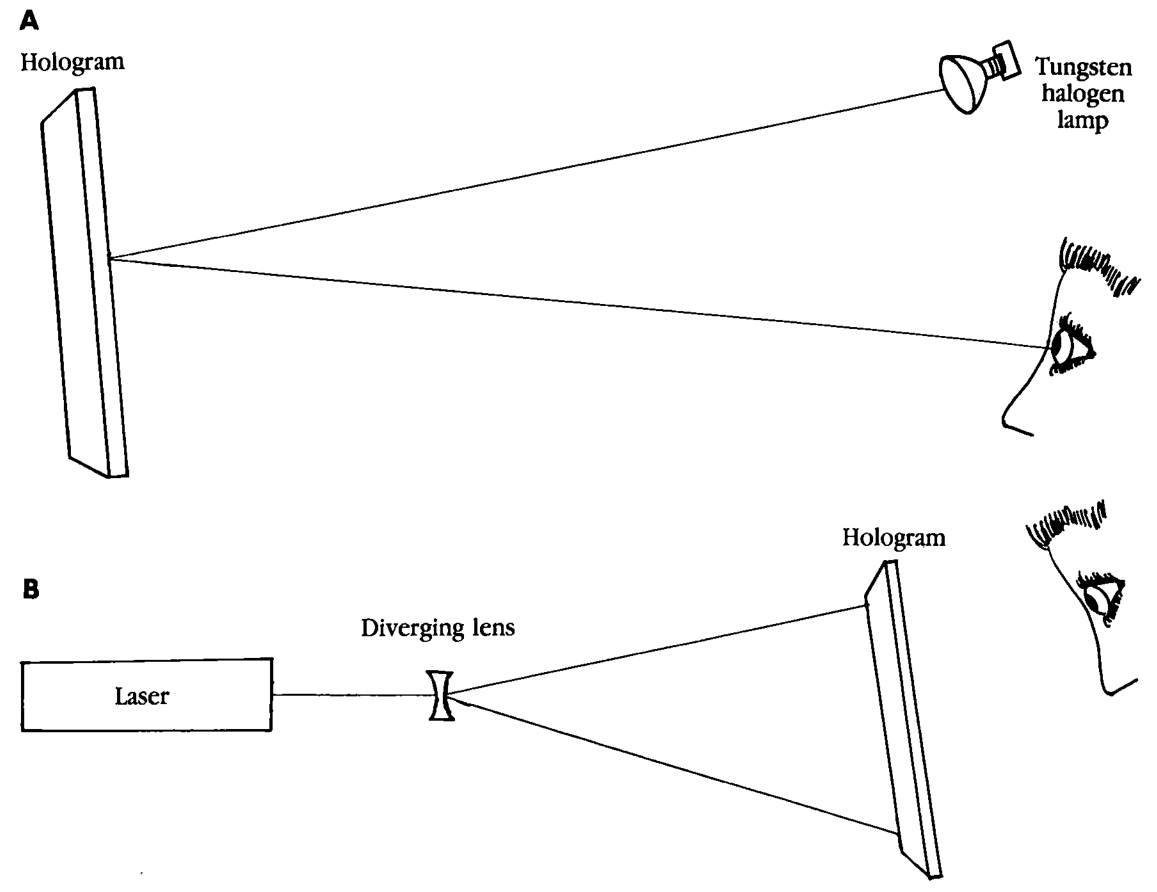

7-12 Viewing holograms: (A) is reflection and (B) is transmission.

Transmission holograms

Transmission holograms need a monochromatic light source for viewing. Use the laser you used to produce the hologram to view the hologram. A filtered white light source can also be used but requires a special narrow-band filter. Edmund Scientific sells this type of filter. They have a bandwidth of ± -2 nm.

I haven't actually used these filters, so I don't know much about them except that they are available. If you purchase a filter for illumination of transmission holograms, get one that is active at around 633 nm, which matches the wavelength of the laser light used to produce it. If they don't have a 633 nm filter, go for a shorter wavelength, which should illuminate the hologram more brightly than a longer wavelength. However, be aware of the filter's heat resistance, which determines how close to the light source you can place the filter.

With transmission holograms, the light must pass through the film, so you observe the hologram on the opposite side of the light source (see FIG. 7-12B). Spread the laser beam to cover the hologram completely. When looking at the hologram, look at the plate, but do not look directly into the laser. Tilt the hologram until you find the best angle for illumination. Remember, a 5 mW laser needs to be spread to a diameter of 8 inches before it's considered safe. Flip the hologram around so you can see the difference between the real and virtual images.

White light transmission holograms

Transmission holograms can be made to be viewable in white light (rainbow holograms). These procedures are explained in chapters 8, 9, and 10.

TROUBLESHOOTING HOLOGRAMS

Not all the holograms you'll shoot will be perfect. The following list is to help you locate the problem when you encounter some fault with the hologram's image.

No image A hologram without any image can be a very frustrating problem. Even though it might seem as if you don't have a starting point to begin evaluating the problem, you do. During development, did the hologram go black or did it gain some density? In the latter case, the most probable cause is that the film plate or model moved during exposure.

Parts of hologram are missing This relates to the first problem. If you are shooting a hologram that contains more than one object, each object must be secure. If an object moves even slightly during exposure, it will not develop a holographic image.

Dark spots on the hologram This is also caused by slight movement of the object. The object could have vibrated or become slightly distorted during exposure.

Image has bands of light and dark The object moved slightly during exposure. If the plate has bands, then it moved slightly during exposure.

Image faint A faint image can be caused by either overexposing or underexposing the film. The best way to check this is by checking the hologram during development against a neutral density filter. Another possible cause is using tap water that contains chlorides in your stock development solutions. Only use distilled or de-ionized water for your stock solutions.

Image is weak and fades This is the problem I encountered when my safelight was fogging my film during setup and development.

Hologram goes dark after a week or two This is called *printout.* It can be caused by residual traces of triethanolamine left in the emulsion.

It's easy to forget basic information when you are excited about shooting a hologram. But the idea is to generate bright, successful holograms, so it's a good idea to keep the following points in mind.

- Always allow sufficient time for your laser to warm up. This is usually about one half hour. As the laser warms up, the glass expands slightly, which varies the longitudinal mode. So it's important that the laser is stable by allowing for this warm-up period before shooting.
- Stability of the optics is very important. If you notice any of the optical assemblies drifting off of their mark, reinforce the assembly by placing another magnet behind it. If you are using the mount directly on the surface of the wood table, use a small amount of clay or Fun-Tak beneath the base plate to prevent the assembly from rocking or vibrating.
- Make sure all your optics are clean. Any dirt or dust will add "noise" to the laser beam that will show up on your hologram.
- Keep it simple. When advancing onto split-beam work described in the following chapters, keep the geometry of the table as simple as possible to accomplish the task at hand.
- In split-beam work, remember to measure your beam path distances starting from the beam splitter. Try to keep the distances as equal as possible without making yourself crazy.
- Haste makes waste. Old but true, allow yourself plenty of time to arrange your setup and shoot. If you can't accomplish it all in one night, stop and pick it up the next evening.

Chapter 8

Advanced techniques

The techniques described in this chapter are illustrated using single-beam holography to allow you a large latitude for experimentation while using simple setups. Inasmuch as split-beam holography improves the resulting image, you can achieve the same positive effect when employing these techniques in your split-beam work.

WHITE LIGHT REFLECTION COPY

As explained in chapter 1, holograms produce both a real and a virtual image. If you flipped a reflection hologram so that you are viewing the real image, you can use it to make a copy, or *transfer hologram.* Because the real image lies "in front" of the plate, you can position the copy plate to intersect the image behind, straddled, or in front of the new hologram, depending on the distance between the plates when making the transfer. You can accomplish the same effect using a transmission hologram, so this same setup is suitable for producing reflection copies from either transmission or reflection masters.

Recall that the real image is pseudoscopic (false), so after exposure and processing, you must flip the transfer hologram to view it. In doing so, you would be looking at a real orthoscopic image. The parallax of the transfer hologram is limited, unless you have used a large master hologram to make the transfer hologram, because when making the transfer hologram, the real image was recorded along with its particular parallax.

Any good quality reflection or transmission hologram you like can be used as a master. Look at FIG. 8-1. The master hologram, hereafter referred to as "H1," is placed in the spread laser beam. The real image is being projected toward the laser. Adjust the angle that the light is incident on the plate to produce the brightest image. Now using a card the exact size of the copy hologram, and position it to be parallel and in front of the master hologram. Position the distance of the plate to intersect the real image where you want it to lie. The shadow cast by the card should fall in line with the edges of the

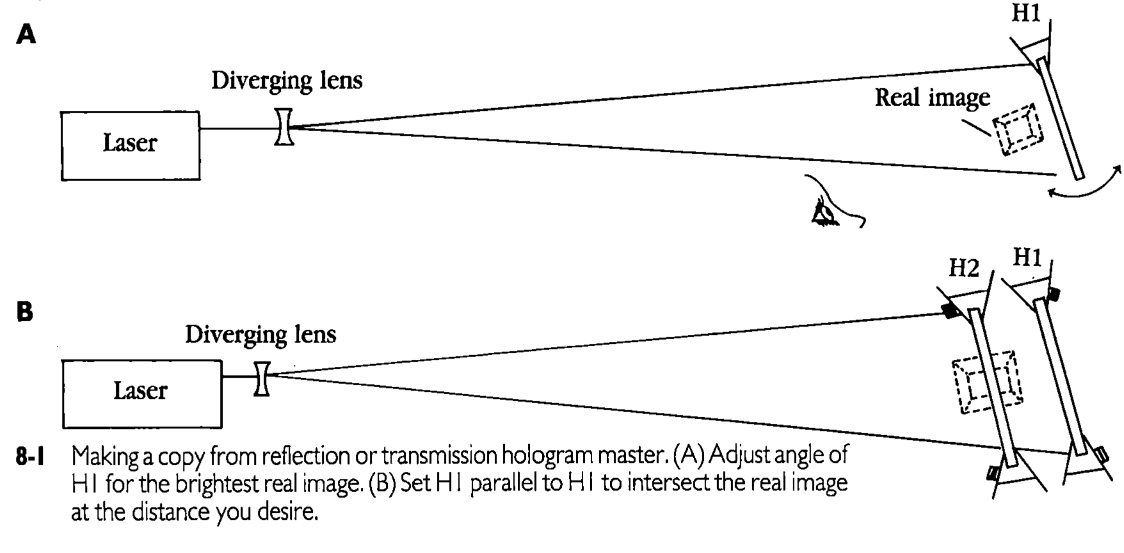

8-1 Making a copy from reflection or transmission hologram master. (A) Adjust angle of H1 for the brightest real image. (B) Set H1 parallel to H1 to intersect the real image at the distance you desire.

master hologram. Position your plate holders at this position and remove the card.

Place a fresh plate, "H2" in the illustration, in the plate holders with the emulsion side facing the H1 master. Make the exposure and develop. Use the same exposure time you used to record the master hologram.

TRANSMISSION COPY

Arrange the geometry of the table as in FIG. 8-2. This setup, when performed correctly, produces a transmission copy that is close to *achromatic* (black and white image) under white light and produces a fine image under laser illumination. Consider it a "cheap and dirty" achromatic white light transmission hologram. Adjust the master hologram to produce the brightest real image. Set up your copy plate to intercept the real image and at the same time have an unobstructed reference beam. Not an easy thing to do, and the

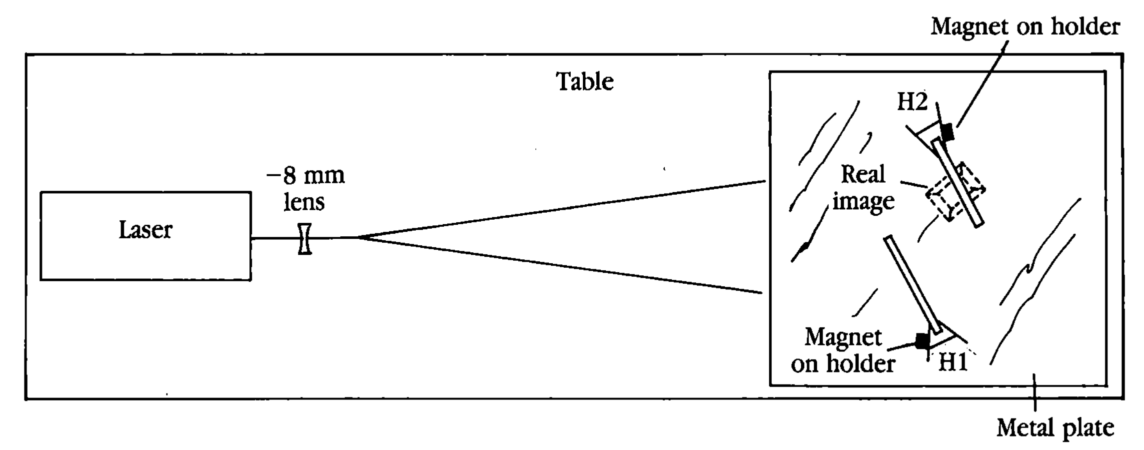

8-2 Making a transmission copy.

geometry would not be perfectly aligned as illustrated. You'll have to make compensations but the results are well worth the endeavor.

Reflection/transmission copy holograms are covered in chapter 9.

DOUBLE-CHANNEL HOLOGRAM

An interesting technique used quite successfully in commercial applications is a *double-* or *dual-channel hologram.* Despite the tremendous amount of information recorded in a single hologram, two or more separate holograms can usually be recorded onto a single plate. Another word for dual-channel is *multiplex.* In dual-channel holograms, the image changes as your viewing angle changes. As an example, one view would be of an unopened jewelry box, and another view would show the same jewelry box opened, revealing an engagement ring. See FIG. 8-3.

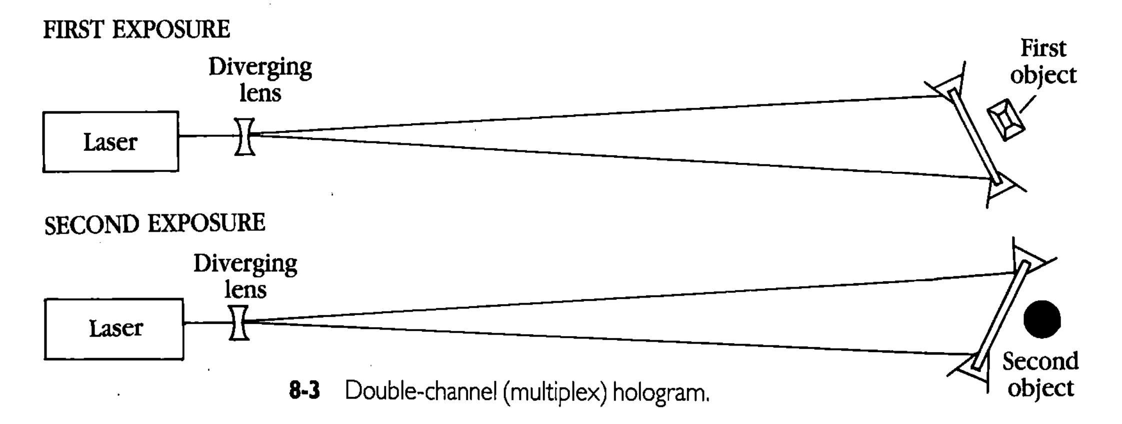

8-3 Double-channel (multiplex) hologram.

The easiest method of recording multiple images on a hologram is to vary the angle of the reference beam on the plate between exposures. The technique can be used with both reflection and transmission holograms. Each exposure should be about half of the typical exposure used to record the hologram. By keeping the difference of the reference beam angle large, there is good separation of the images with little cross talk between them.

PSEUDOCOLOR REFLECTION HOLOGRAM

Holograms typically play back their images in the same monochromatic red they were recorded in. To produce color holograms, a number of methods are available. One method requires two or three lasers, each in a different primary color. Aside from the high expense of the lasers, this involves accurately mixing the laser light together at the correct power levels; the emulsion material is not equally sensitive at all frequencies. Despite the cost and difficulty, full-color and naturally colored holograms have been produced this way.

White light transmission holograms (rainbow holograms) can also be used to generate full-color holograms, but it involves complex table geometries that aren't suitable for beginners.

Pseudocolor reflection holograms can be made with an HeNe laser by a technique that swells the emulsion plate with triethanolamine (TEA) before plate exposure. After the TEA-swelled plate is exposed, the interference pattern recorded in the plate as usual. During development, the TEA washes out of the emulsion, which shrinks the emulsion back to its normal size. But the interference pattern is tighter, so the hologram plays back at a shorter wavelength (see more in chapter 13).

Different concentrations of TEA can provide reflection hologram playback in yellow, green, or blue. The method is well suited to single-beam work. TEA has a hypersensitizing effect on the emulsion, depending on its concentration. Reduce the exposure time for TEA-treated plates by a factor of two or three. TEA also begins darkening the emulsion material after about 4 hours without any exposure to the plate, so do not try to make up a batch of TEA plates for eventual use. For the same reason, don't place any of your finished bleach holograms in TEA to shift their wavelength.

Triethanolamine is a clear fluid that dissolves in water. It as available from the Photographers Formulary (see "Sources"). Try the following concentrations (use as a guideline).

0% red
10% yellow
14% green
20% blue

To make any concentration, match the percent required in milliliters of TEA. Add to this distilled water to obtain 100 milliliters total. Therefore, for a 10 percent solution, you would measure 10 milliliters of TEA and add 90 milliliters of distilled water.

Under a safelight, pour the TEA solution into a development tray. Put an unexposed plate into the tray, emulsion side up. Let the plate sit in the TEA solution for about a minute. Remove the plate and squeegee the excessive solution from the plate. Allow the plate to dry completely (about 20 minutes).

Expose the plate using a shorter exposure time than normal; try about half the standard exposure time. Develop and process the plate normally. When the plate dries, check the color shift of the hologram. The percentages of the solution required to bring about a particular color shift vary with the batch of emulsion plates used and as well as the method of swelling and drying the plate.

This is a fertile area for experimentation. Try making a two-channel, two-color reflection hologram. To make a pseudocolor hologram, combine the techniques presented here with the two-channel information. Techniques like this have already been employed with one major difference. In the two-channel holograms, the reference beam angles greatly differed between each channel. To create a pseudocolor hologram, the reference beam is only diverted about 7 degrees between each exposure. This creates cross talk between the images, so they both are visible simultaneously.

If you want to try a three-color pseudohologram, use the following exposure times:

Red ½ standard exposure
Green ½ red exposure
Blue ½ green exposure

Between each exposure, change or move the object you are holographing.

I advise making single-color holograms first to determine the proper TEA concentrations for each color. Then proceed to two- and three-color holograms.

Consider these experiments to be the tip of the iceberg. There are a number of ways to introduce color into holograms. Some use transmission masters. You can get more accurate color registration by reading the article in the below list by S. McGrew on geometry. If you are interested in pursuing this line of holography, read or obtain the following papers.

- Moore, L. "Pseudocolor Reflection Holography." Proceedings of the International Symposium on Display Holography (1982).
- McGrew, S. "A Graphical Method for Calculating Pseudocolor Hologram Recording Geometries." Proceedings of the International Symposium on Display Holography (1982).
- St. Cyr, S. "One-Step Pseudocolor WLT Camera for Artists." Proceedings of the International Symposium on Display Holography (1985).
- Smith, S. "Applications of Tri-color Theory of Additive Color Mixing to the Full-Color Reflection Hologram." Proc. of SPIE, vol. 523 (1985).

If a library in your area doesn't have these papers, the first three are available from the Holography Workshops, and the fourth is available from SPIE (see "Sources").

Chapter 9

Split-beam holography

Splitting the laser beam provides greater control over the ratio of light intensity that reaches the plate from the reference and object beams. In addition, the illumination of the object can be more creatively applied. With these benefits come certain responsibilities. First, the length of the individual beam paths to the holographic film plate should be made equal to one another. The beam ratio is determined to reduce the amount of noise generated in the hologram. To understand how this functions let's first look at how noise is generated.

NOISE

Noise can be created by intermodulation of the laser light off of the object (see FIG. 9-1). A hologram is a recording of interference patterns created by the reference and object beams. However, the light reflected off of an object can create an interference pattern that is independent of the reference beam. In the figure, as the reflected light from two arbitrary points A and B propagates towards the plate, the light creates its own interference pattern that is recorded in the film emulsion. The larger the object and the closer to the plate, the more pronounced the effect. This interference created by two points on the object is called *intermodulation.*

Intermodulation angle

There are two methods to reduce the intermodulation effect. First, make the angle between the object and reference beam greater than the intermodulation angle between A and B. The angle created between points A and B is less than the angle formed by either A and R or B and R. The smaller angle ensures that the minimum spatial frequency between the object and reference beam is greater than the spatial frequency of the noise created by A and B. Spatial frequency relates to the lines per millimeter recorded on the film plate. A higher spatial frequency means that it forms more lines per millimeter.

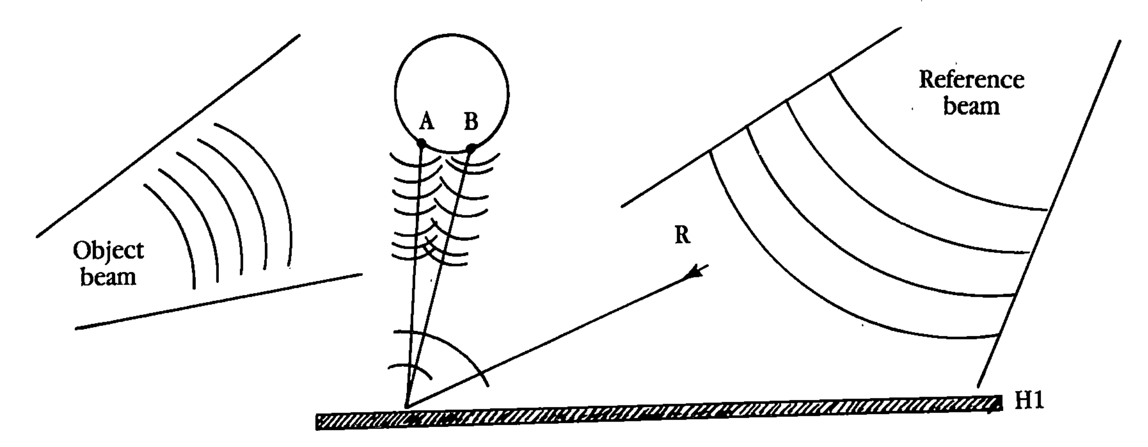

9-1 Creation of intermodulation noise.

Beam ratio

A second way to reduce intermodulation is to use a reference beam that is brighter than the object beam to ensure that the interference pattern between any two points from the object are weak in comparison to the pattern formed by the object and reference beams. As a general rule of thumb, the ratio of light for a reflection hologram should be approximately 1.5 to 1 with the reference beam the brighter of the two beams. In making transmission holograms, use a ratio of 4 to 1, again with the reference beam the brighter of the two beams. These ratios are estimates, and depending upon the setup can vary widely from the stated ratios. For instance, for a transmission hologram of an object that filled a wide angle of view (as seen from the film plate), a better ratio of light would be 8 to 1. This ratio would help to eliminate noise in the hologram. I suggest that you take a somewhat cavalier attitude towards the beam ratio. Use the estimates 1.5:1 for reflection and 4:1 for transmission, and don't vary them unless you start getting poor results. At that point, you'll have to discover the proper beam ratio by trial and error (which isn't as bad as it might seem). The beam ratios are not written in stone, and they are not that critical. If a particular setup says that it requires a 4:1 beam ratio, a 3:1 or 5:1 ratio usually works just as well.

Estimating beam ratios Beam ratios are adjusted primarily with the beam splitter. Beam splitters, as stated previously, come in a number of fixed ratios such as 1:1, 2:1, 3:1, 4:1, and so on. More expensive variable-ratio beam splitters are available (they are not required for the projects in this book). These units have various increments of beam ratios stamped on their surface. The location of the beam through the surface determines the ratio.

LIGHT INTENSITY

After the laser has been split by the beam splitter, you can still adjust the light intensity by varying the distance of the expansion optic from the plate or object, which controls the beam spread, and the further you spread the beam, the less intense the light (see Appendix A).

When you first start shooting split-beam setups, you can use your eyes to estimate the beam strength. Place a white card the size of the plate you are shooting in the position you will be putting the film plate. Block the light from the reference beam and observe the light intensity falling on the white card from the reflected light off the object. Now block the object beam and check the light intensity of the reference beam. Adjust accordingly.

Photometers are available that measure the intensity of the light. You can purchase an inexpensive photometer from Edmund for about $150, or you can build one for about $30. Plans for building a photometer are in Appendix B. The homemade photometer can measure the direct power output of your laser and check the light ratios in your split-beam setups. Whether you purchase a unit or build one, the methods for use are the same.

All photometers have a light sensor. Place the light sensor in the path of the beam. Block one beam and measure the light intensity of the other. The photometer is calibrated in mW (milliwatt) or μW (microwatt) per square centimeter.

When the reference beam of the laser is used to reconstruct the image in a hologram, it is called the *reconstruction beam.* When you flip the hologram around to view or project the real image, the reference beam is now traveling through the hologram in the opposite direction from which the hologram was made. This reconstruction beam is said to be the conjugate to the reference beam and is labeled the *conjugate reconstruction beam.*

SIMPLE SPLIT-BEAM REFLECTION HOLOGRAM

The simple split-beam setups using the 2.5-square-inch film plates easily fit on the small table. More complex setups could be squeezed onto the table, but in these cases, you are better off working on the more advanced table described in the next chapter. When working with 4 × 5-inch plates, the space available on the small table is also probably inadequate. You might be able to fit very simple split-beam setups on the table using the larger film plates, but more complex setups require the larger table.

Figure 9-2 illustrates a simple split-beam setup. Notice the −8mm lens is used to spread the object beam. You can also use a spherical mirror to spread the object beam as described in the single-beam set-ups (chapter 8). Notice that

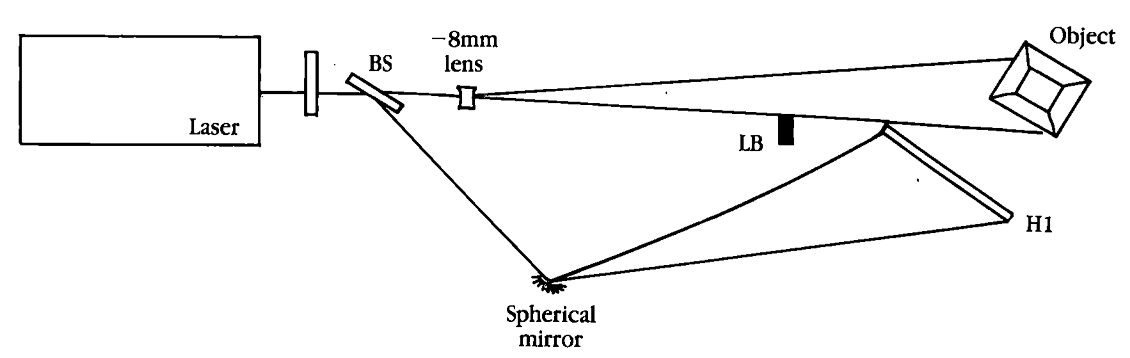

9-2 Simple split-beam reflection.

the reference beam also provides illumination to the object, providing frontal illumination from the reference beam and side illumination from the object beam. Although this provides even lighting of the object, you must keep the object placement with the one-half coherence length for the reference beam to keep the reflected light coherent (see the section on single-beam reflection holograms in chapter 8 for more complete information). So the disadvantage of this setup is a limited depth of field.

Improved simple split-beam reflection hologram

Figure 9-3 is an improved set-up from that in FIG. 9-2. Notice the reference beam is aligned so that it doesn't illuminate the object. Although the lighting is uneven due to the shadow cast by the object beam, the depth of field increased more than two times that of the single-beam reflection setup.

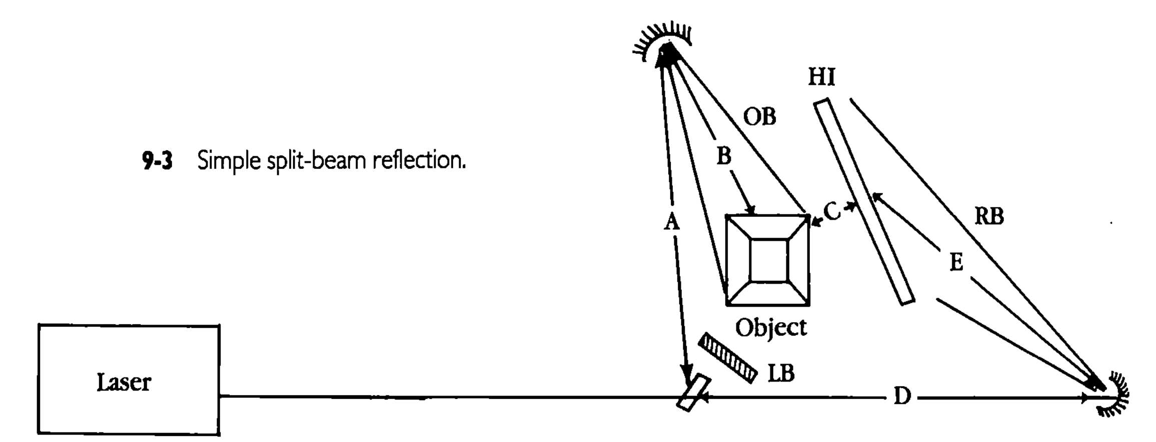

9-3 Simple split-beam reflection.

Equalize path lengths

In all split-beam work, it's important that you keep the lengths of the split-beam paths equal to provide maximum interference and depth of field. In FIG. 9-3, this means the lengths of A + B + C (the object beam) should equal the distance (or length) of D + E (the reference beam). Note in particular the distance C, the reflected light from the object to the plate. This is the most commonly missed distance when measuring and matching beam lengths. Don't got too bogged down with matching the beam paths. It isn't necessary to have the paths matched to exactly 0.001 inch; just get them as close as possible, within an inch or two, without becoming obsessed. Save your obsession for the ultimate hologram. Notice the setup uses the spherical mirrors exclusively, which allows for the use of short beam paths (easy to match) that easily fit on one end of the table.

Check the beam intensity ratio

In this set up, the beam splitter is 1:1. So ideally, both beams are starting out with the same brightness from the beam splitter. Check the light intensity, using your eyes or a photometer (explained previously in this chapter).

In this simple setup, the reference beam is more than likely much brighter than the object beam because in the latter case, the light is reflected off the object. With spherical mirrors, you can't simply move the object mirror closer to the object to increase its illumination because you would change the beam path length. You can circumvent this situation by using two transfer mirrors as illustrated in FIG. 9-4.

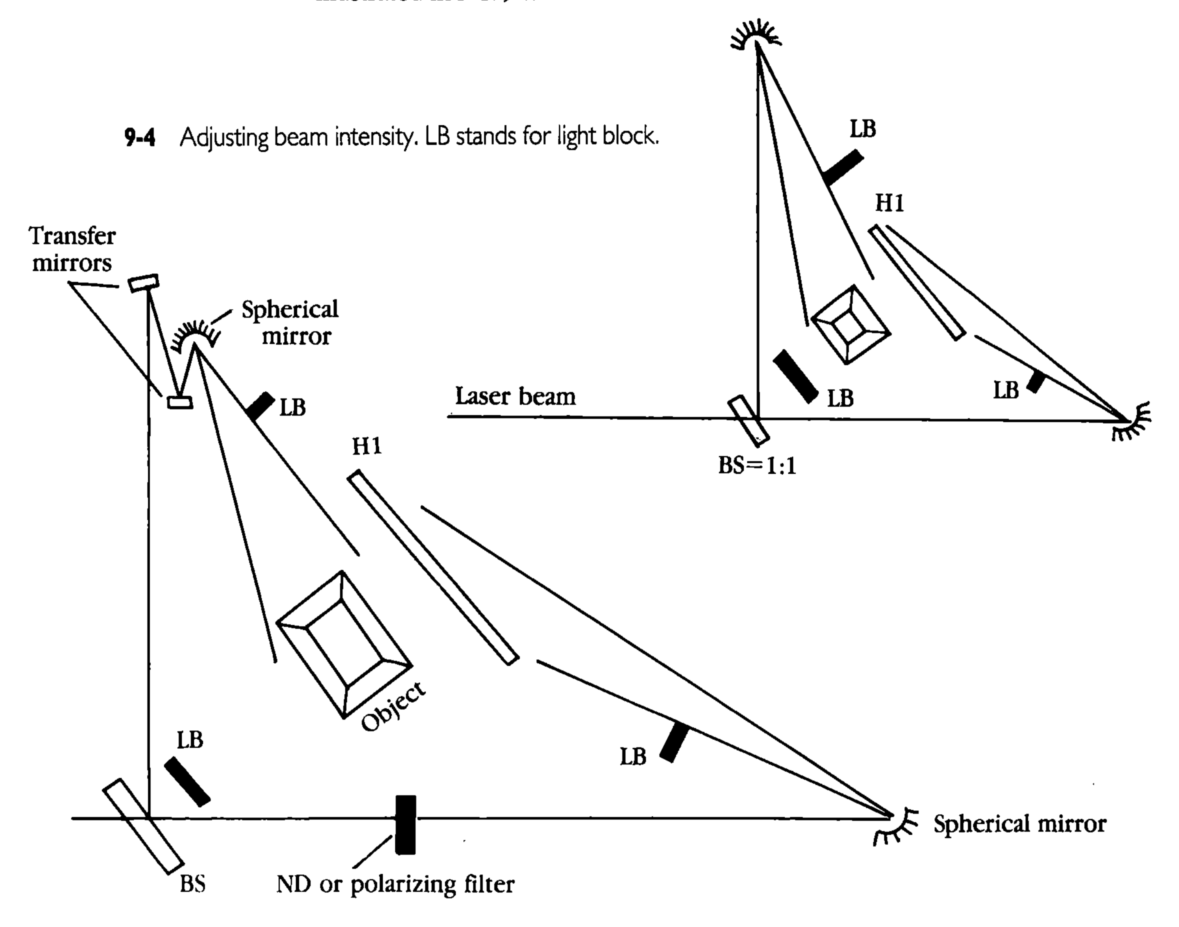

9-4 Adjusting beam intensity. LB stands for light block.

Another possibility is to use a small piece of the 0.3 neutral density filter in the path of the reference beam to reduce the light intensity and make it more evenly matched to the object beam. Still another possibility is to use a small piece of polarizing film in the reference beam to lower its intensity.

The diagrams of the table geometry are working models. They are not "written in stone." Don't be afraid to change the geometry to suit your needs. As an example, FIG. 9-5 illustrates essentially the same setup as FIG. 9-4, except the geometry is adjusted to provide better lighting for the object.

Figure 9-6 is a typical split-beam reflection setup. Beam paths are measured starting from the first beam splitter. You now need to match three beam paths instead of only two.

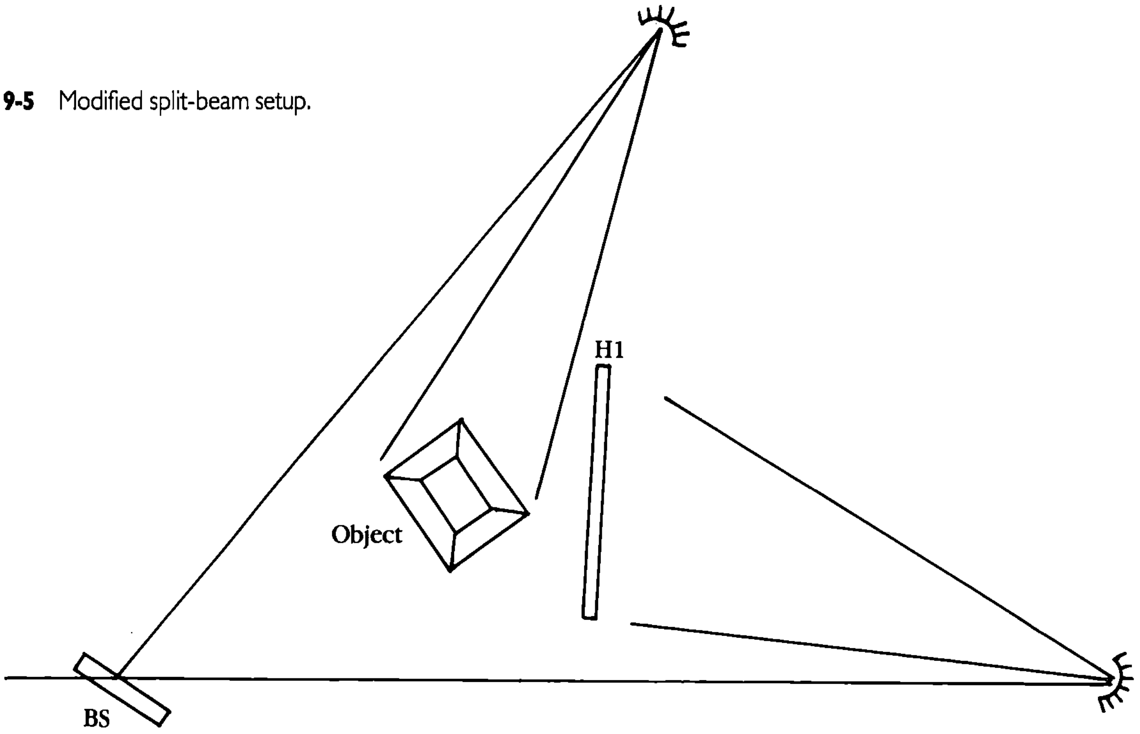

9-5 Modified split-beam setup.

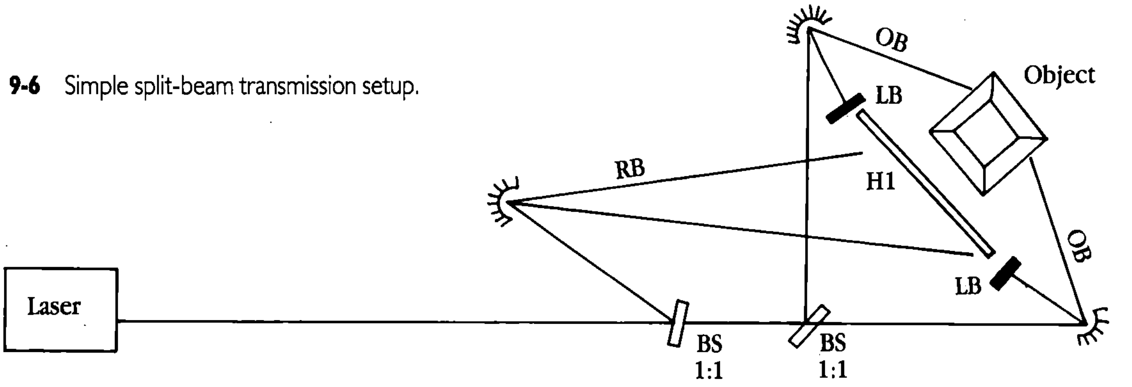

9-6 Simple split-beam transmission setup.

SIMPLE SPLIT-BEAM TRANSMISSION HOLOGRAM

Figure 9-7 illustrates a simple split-beam transmission setup with a 3:1 or 9:1 beam splitter. The procedure is the same as with the split-beam reflection: match the beam paths, and check the light intensities before shooting. With the transmission hologram, you want to achieve a 4:1 beam ratio. Remember that the reference beam (the transmitted beam from the splitter) is the brighter of the two beams.

Figure 9-8 is a typical split-beam transmission setup. Notice how similar it is to FIG. 9-6.

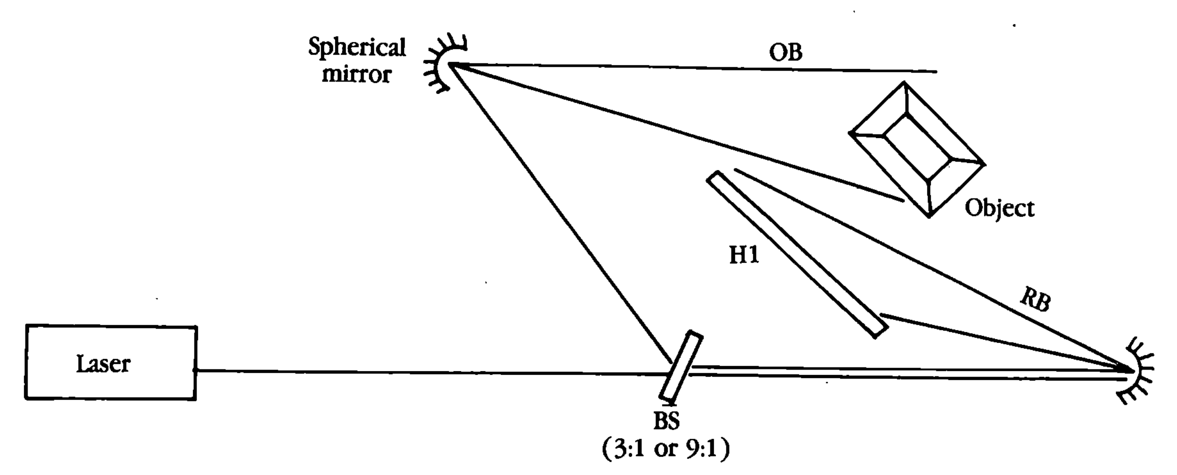

9-7 Split-beam reflection setup.

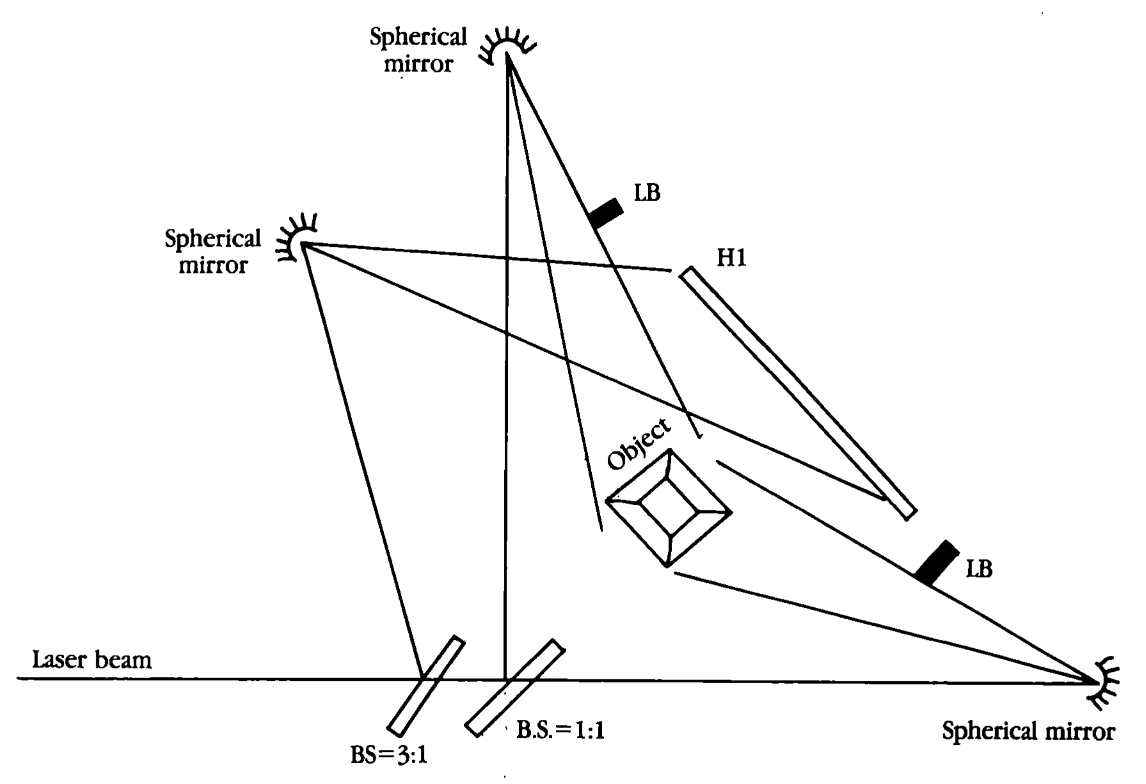

9-8 Split-beam transmission setup.

IMPROVING THE QUALITY OF SPLIT-BEAM SETUPS

Some of the various ways you can improve split-beam shooting are using a polarized laser, a half-wave plate, a spatial fitter, or a collimating lens.

Polarization

If you followed by suggestion and are using a polarized laser, you can capitalize on it now. Its important for the reference beam and the object beam to be in the same plane of polarization (see chapter 11). If the planes match, strong interference patterns form that are recorded in the film. If they are not matched, weaker interference lines form, their strength dependent on the angle difference of the planes. At 90 degrees apart, no interference patterns will form.

When the laser light reflects off of the object, the light is usually depolarized. Only the light that remains in the same plane as the reference beam contributes to forming the hologram. The balance of the light is useless and will fog the film. A half-wave plate is a simple optical element that can align the reference beam's polarization to maximize the interference with the object beam.

Half-wave plate

The half-wave plate, also known as *retardation plates,* are usually made of a thin sheet of quartz or mica. These plates are available in quarter-wave, half-wave and full-wave sizes. The plate material retards the light passing through it: half-wave retardation plate slows the beam one-half wave. You rotate the plate to control the retardation and phase of the light, which also rotates the plane of polarization.

To align a half-wave plate, place it in the reference beam path. Block off the reference beam. Using the polarizing material, rotate the polarizing material while looking at the object from the plane of the film plate. Mark the position of the polarizing material when the image is at it brightest. Now block the object beam and unblock the reference beam. Place a white card at the film location. Position the polarizing material in front of the card so that the reference beam casts its shadow on the card. The polarizing material must be in the same orientation that was marked when you checked the object beam. Rotate the half-wave plate until the light passing through the filter is at its brightest. The beams are now matched to produce the strongest interference pattern.

Making a half-wave plate I haven't seen any half-wave plates on the surplus market, but I have seen quarter-wave plates. Two quarter-wave plates aligned together can act as a half-wave plate. Meredith Instruments currently sells quarter-wave plates for $5 each. I purchased two plates to make up one half-wave plate.

If you are buying on the surplus market, it is nearly impossible to determine exactly what configuration the quarter-wave plate will be in. Figure 9-9 is a photo of a quarter-wave plate I purchased. The plate is mounted in plastic with one end of the plastic rotating the plate. The plastic is held with a snap ring to a metal L bracket. I removed the snap ring with needle-nose pliers and separated the plastic from the metal bracket (FIG. 9-10). At this point, check the plastic holder to see which end rotates with the plate inside. The end that does not rotate the plate is glued to a rectangular magnet. Making sure the hole in the magnet is aligned with the opening in the quarter-wave plate. Another quarter wave is mounted to the other side of the rectangular magnet to make a

9-9 Quarter-wave plate as received

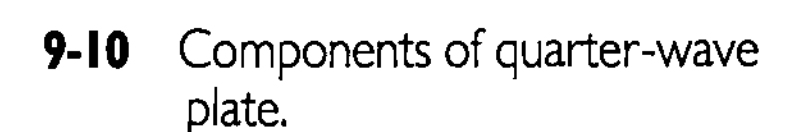
9-10 Components of quarter-wave plate.

half-wave assembly plate. This half-wave plate is mounted like any other optical component (see FIG. 9-11). To use, align the plate in the reference beam, and you can rotate either side of the plate to adjust the plane of polarization. For the sake of illustration, the half-wave plate should be inserted in a setup like the filter shown in FIG. 9-4.

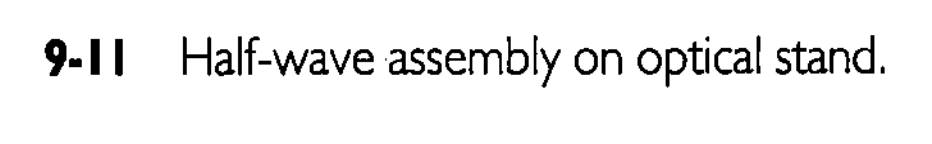
9-11 Half-wave assembly on optical stand.

Spatial filter

Spatial filters are used to "clean up" the beams. If you only have one spatial filter available, use it to clarify the reference beam because that would carry the most impact on the quality of the resulting hologram. Use the simple spatial filter project in Appendix B in the simple split-beam setups using the 2.5-square-inch plates.

Collimating lens

To *collimate* light means to project plane waves. The light emanating from simple lenses or spherical mirrors is diverging (spreading outward). Large lenses or mirrors can be used to collimate a laser beam (see FIG. 9-12). You place the diverging optic at the focal length of the collimating optic. The area the diverging lens illuminates up to the collimating optic determines the size of the resulting beam.

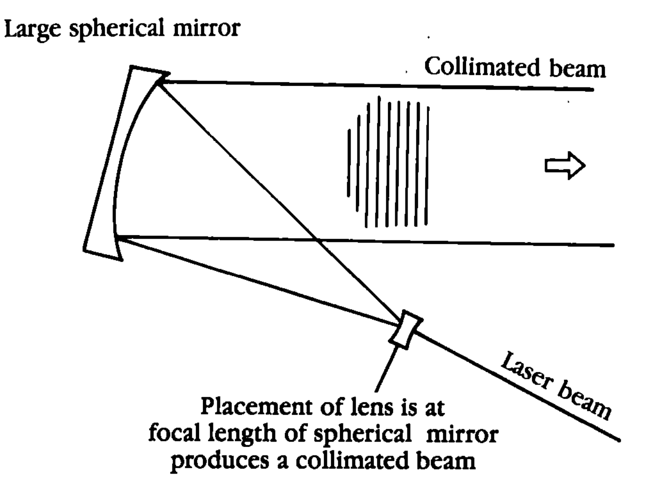

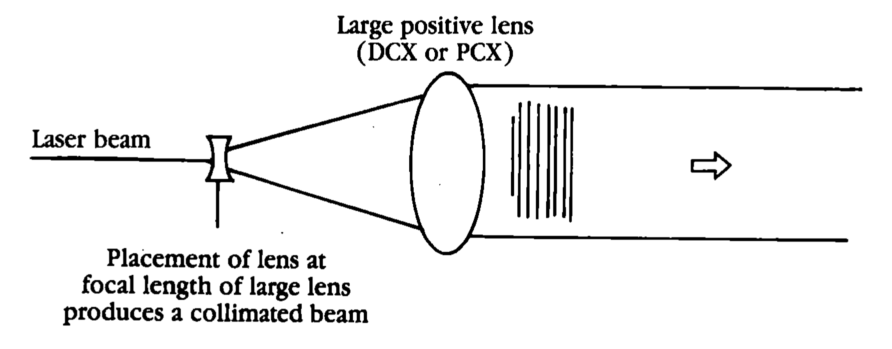

9-12 Using a mirror or lens to collimate laser beam.

Typically, the reference beam is collimated so that the real image can be projected with less distortion and swing. This becomes critical when you want to generate high-quality transfer holograms. Large lenses are typically more available and cheaper than large, front-surface spherical mirrors. In addition, they can be used quite easily with the common small spherical mirrors, ensuring an adequate beam spread up to the optic and a good-sized collimating beam.

The transfer illustrations that follow show collimated object beams. If the master hologram was produced with a collimated reference beam, you need to use a collimated beam to illuminate the hologram properly; in the transfer process, it is the object beam. You can produce transfer holograms using the diverging optics, but when you want to improve the quality of the transfers, you need to collimate the reference beam.

COPYING HOLOGRAMS

Master holograms are so named because they are made to produce duplicates (*transfers*). Such holograms are produced identically to the way all of the previous ones have been made with one important exception: the reference beam is collimated. When using a collimated reference beam, its important that the beam doesn't "splash" onto the object you're trying to holograph. You can use an overhead reference beam to accomplish this (bounce the beam off of an overhead mirror). Another way is to lay the hologram on its side to allow you to use a side reference beam that appears to the hologram as an overhead beam.

Figure 9-13 illustrates how to make a reflection copy from a reflection master. As with the single-beam copy process and all other transfers to follow, you adjust the master to produce the brightest real image. Figure 9-14 illustrates making a reflection copy from a transmission master. Figure 9-15 illustrates a restricted-aperture reflection copy from a transmission master. The aperture

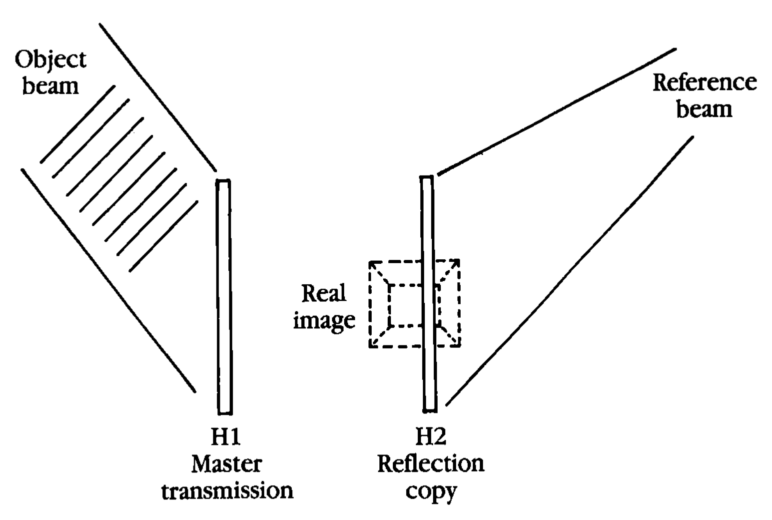

9-13 Making a reflection copy from a reflection master.

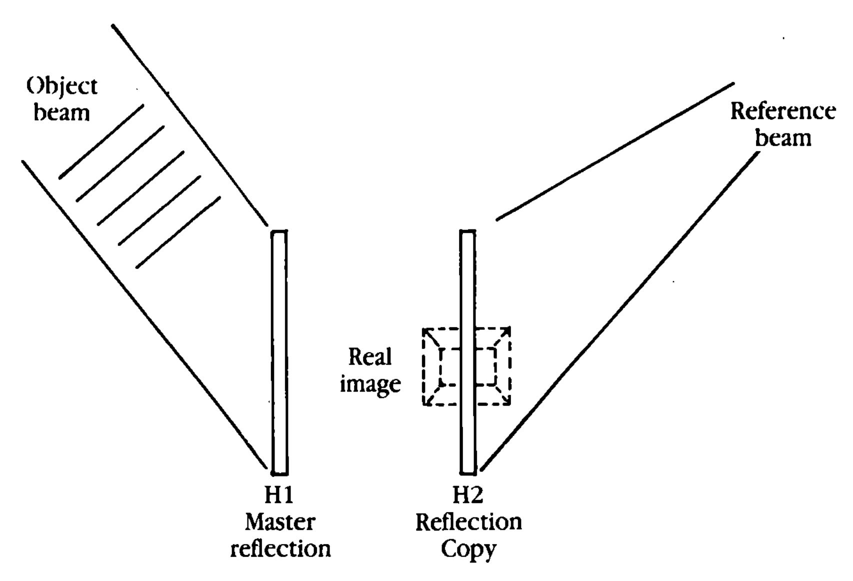

9-14 Making a reflection copy from transmission master.

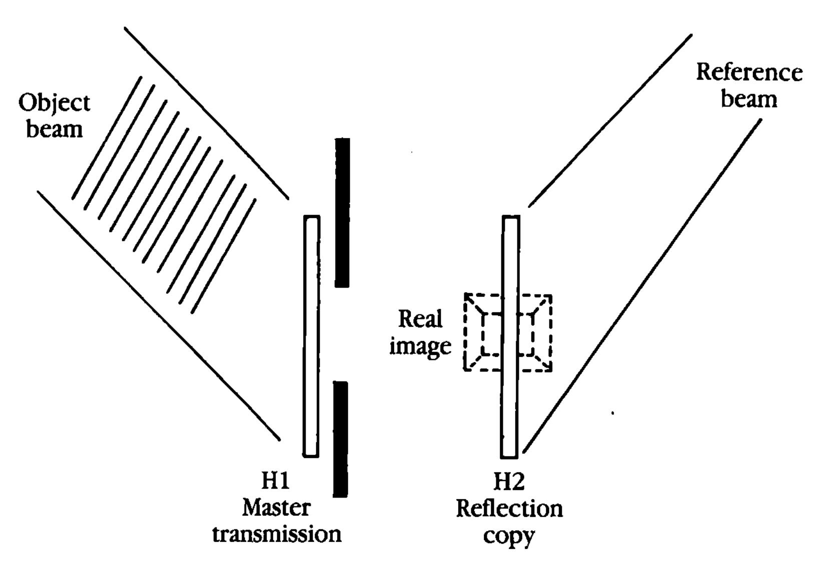

9-15 Aperture restricted reflection transfer.

only allows about half of the transmission hologram to project the real image, which results in a much brighter image for the copy.

Figure 9-16 illustrates the basic configuration of a white light transmission hologram from a transmission master.

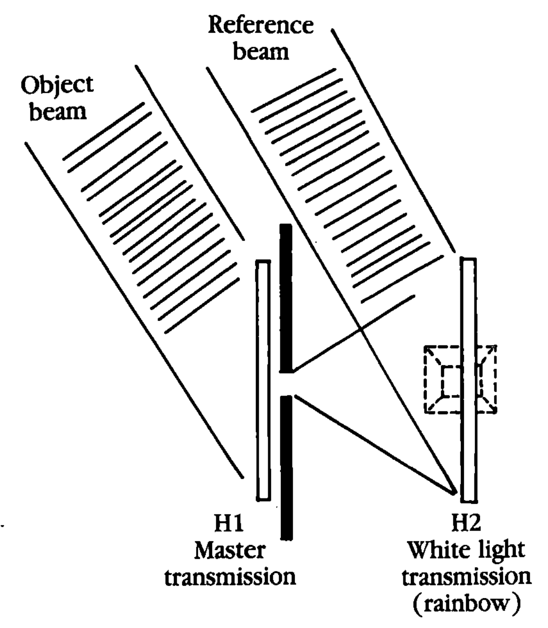

9-16 White light transmission transfer.

Chapter 10

Advanced projects

This chapter contains miscellaneous projects and techniques to help you expand your holography skills. Included are the larger isolation table, making rainbow holograms, single-step transmission holograms, stereograms, using an LCD display or film, and notes on red-sensitive DCG and pulse laser holography.

ADVANCED TABLE

Sooner or later, you will probably want or need more working space for your setups. Construction of the simple table only required half of the 8-foot plank. The leftover piece can serve to build a second table just like the first table, and the original table and the new one are subassemblies for a larger table. Because you are more likely to encounter vibration problems with a larger table, you should purchase the Sorbathane pads (Images Co.) if you haven't already.

Figure 10-1 illustrates the setup of the advanced table. The top of the table is made of ¾-inch or thicker plywood. Be sure to purchase a good-quality lumber that is flat. Any bowing in the plywood will make it difficult for it to seat subassemblies properly. You might consider using two Sorbathane pads at each position, one stacked on top of the other, to make the pad thicker and allow more room for play. When the table is finished, check for vibration by building the interferometer as described for the simple table in chapter 2.

Using optical mounts on plywood

When using optical mounts directly on a plywood surface, it is a good idea to secure the base of the mount to the table with a little clay or Fun-Tak to prevent the mount from vibrating. For example, when using the inverted T mount, place a small amount of clay at each end of the base plate and push it onto the table at the location you want. Finish by positioning the vertical sections and whatever optical components you are using.

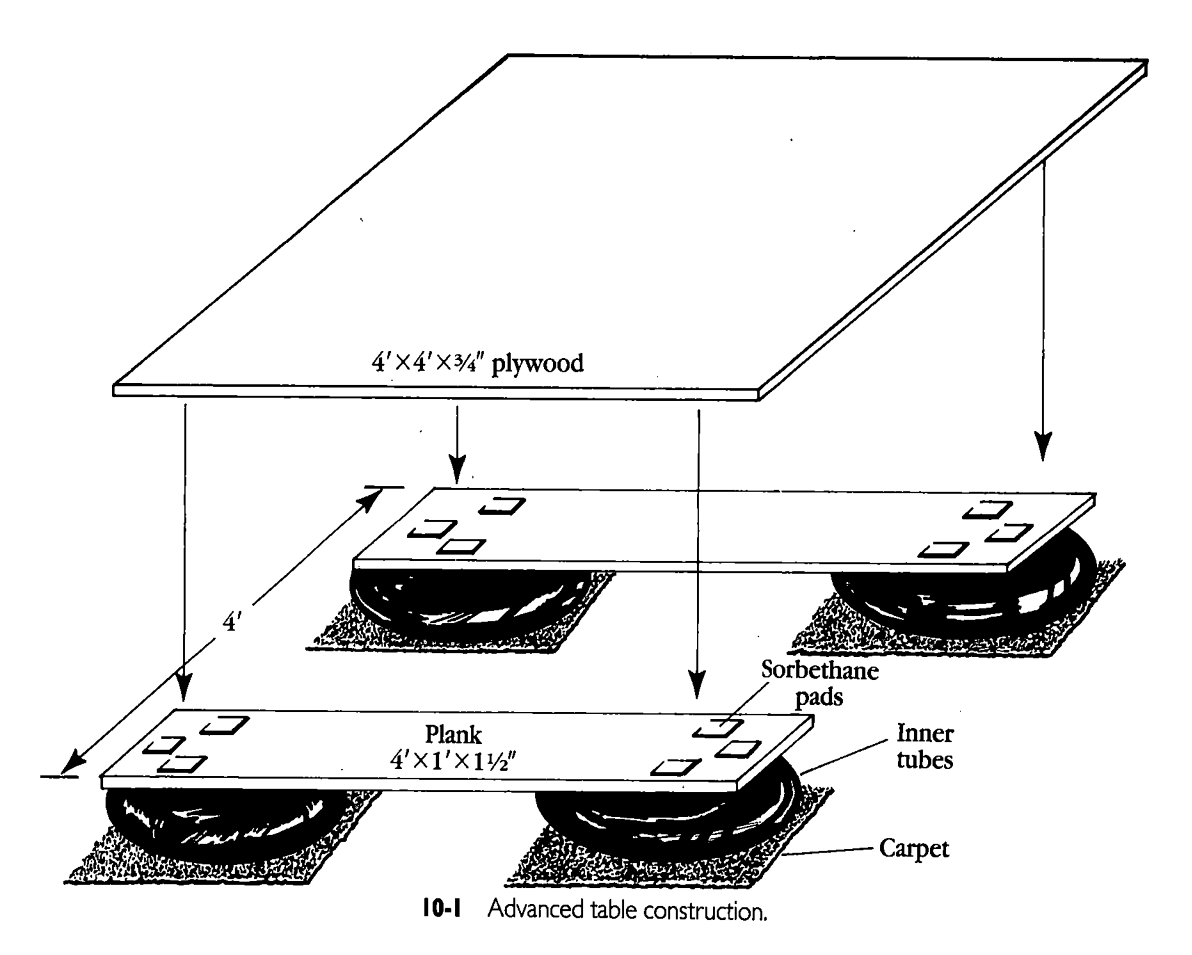

10-1 Advanced table construction.

The large table provides much more room for you to continue producing holograms. The following chapters detail the basic table setups for some advanced hologram shooting. These setups might not be as easy as they look. It's a good idea to get as much background information as possible. The "Sources" section at the end of this book provides addresses and institutions that carry more detailed information on advanced holography setups. Also see the suggested further reading at the end of this chapter.

RAINBOW HOLOGRAM

The rainbow hologram, also known as the white light transmission hologram, was developed in 1969 by Stephen A. Benton of the Polaroid Corporation. At the time, Benton was looking for a way to reduce the information content of holograms for possible TV transmission. Figure 10-2A illustrates what happens when looking at a standard transmission hologram in white light. The observer sees a fuzzy, multicolored image. Benton covered the hologram except for a slit (aperture), as in FIG. 10-2B, and made a copy. The aperture hologram sacrificed most of its vertical parallax but kept the horizontal parallax intact. Since most people typically observe 3-D effects with binocular vision by

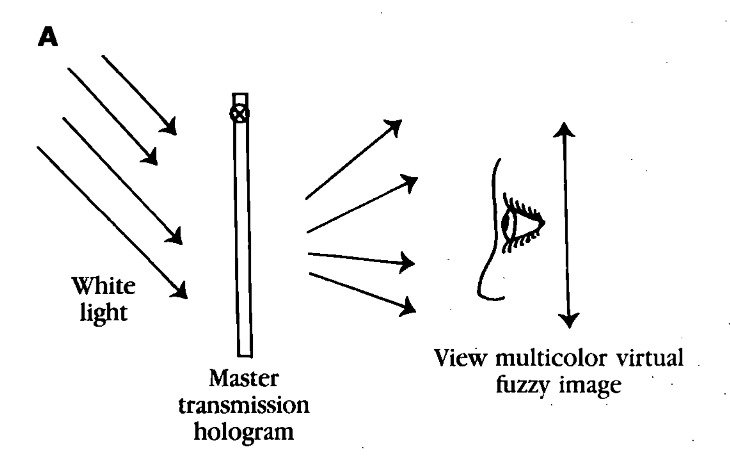

10-2 Making a rainbow hologram from a transmission master.

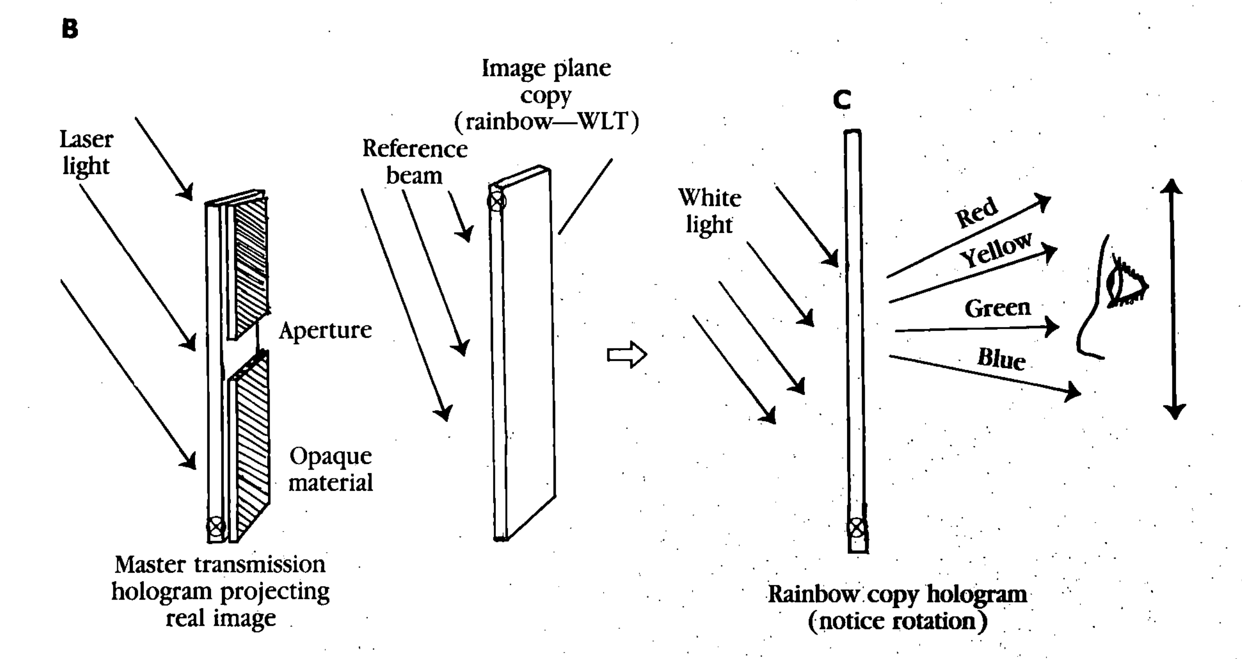

moving the head side to side rather than up and down, the impact of losing the vertical parallax wasn't critical. The advantage of this new copy hologram is that it's viewable with white light, as shown in FIG. 10-2C. As you raise and lower your head while observing the hologram, you see the same image shifting through various colors of the rainbow; hence, the name *rainbow* hologram.

The technique of copying transmission masters is also used to create achromatic hologram images and full-color hologram images. Remember that master holograms are produced using a collimated reference beam. If you want to try shooting a rainbow hologram, keep these points in mind: you are shooting the real image from the master transmission hologram, and secondly, the object beam illuminating the master transmission should not fall directly onto the copy plate. Because the illumination provided from the aperture is dim, you might need to reduce the power of the reference beam. You can accomplish this by spreading the beam further or perhaps placing a 0.5 ND filter in the reference beam path. The width of the aperture should be around 8 to 12 mm, or about ½ inch.

SINGLE-STEP WHITE LIGHT TRANSMISSION HOLOGRAM

Figure 10-3 shows a typical setup that produces a white light transmission hologram right on the table while making a transfer. The aperture is kept in close proximity to the lens. The focal length between the lens and object and the lens and plate is 2FL to produce an image at the film plate that is the same size as the object being holographed.

STEREOGRAMS

Stereograms go under a number of names such as Cross hologram, Benton stereogram, integral hologram, integram and multiplex hologram. Robert Pole of IBM research center in NY created a fly-eye lens that consists of many separated lenses, each of which projects a slightly different angle of a scene onto photographic film. After processing, the film was illuminated with laser light and placed in front of second fly-eye lens system. The wavefront projected by this second lens system resembled the original wavefront and was holographed in the usual manner. This might seem like a lot of back-and-forth movement to make a hologram, but it allows you to produce a hologram that would otherwise be impossible. Substantial improvements in synthesizing holograms from film have been produced by D. J. DeBitettro, Lloyd Cross, and Steve Benton.

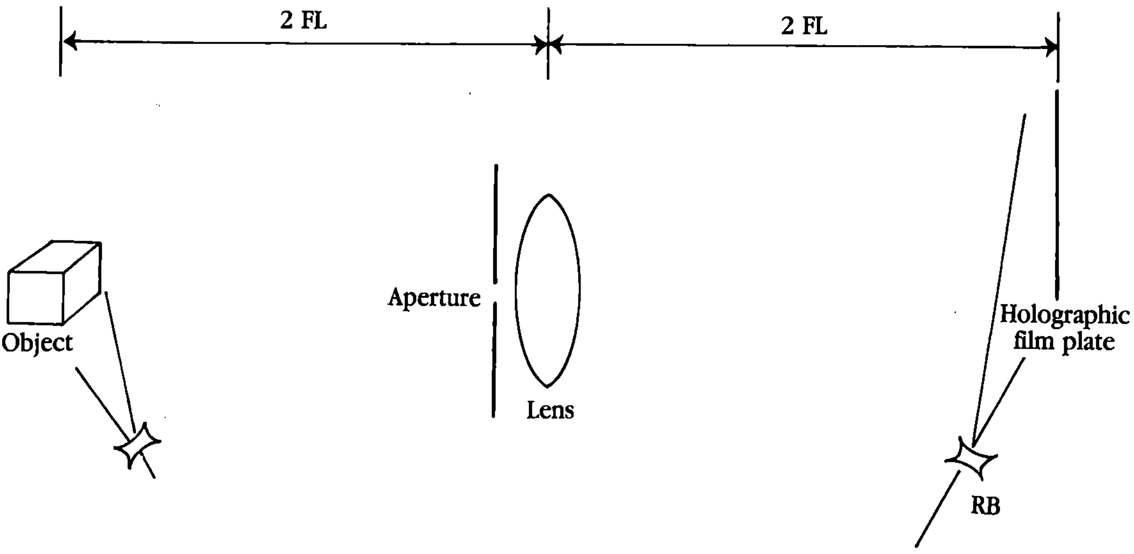

10-3 Single-step white light transmission hologram.

The basic principle to produce a stereogram is as follows. You need a standard 35 mm camera loaded with either color or black-and-white slide film. As shown in FIG. 10-4A, place a subject in front of the camera. Position the distance between the subject and camera so that you will have a reasonable parallax view of the subject through a 2-foot horizontal movement. Shoot one picture for every horizontal inch of travel. Keep the camera level and the subject motionless throughout the picture taking. When you are finished, you will have 24 transparencies of the subject, each from a slightly different angle.

You can synthesize these transparencies into a hologram. See Fig. 10-4B. To create the hologram, place a stationary aperture in front of the film plate. The width of the aperture should be about 3 mm. The lens in front of the transparency focuses the image onto the screen. The screen can be made from a white sheet of tracing paper taped to a piece of glass to keep it rigid. Load the first transparency into the setup. Make an exposure onto the film plate. Remove the first transparency, and load the second one in its place. Move the film plate horizontally a distance equal to the aperture width, in this case 3 mm. Now make the second exposure. Replace the second transparency with the third, move the plate again horizontally 3 mm and make the third exposure. Continue in this manner until all 24 transparencies are holographed, then develop the film plate. You now have a transmission stereogram.

The resolution of this synthesized hologram isn't the best it could be because professional stereograms use hundreds and sometimes thousands of frames to produce one stereogram. These stereograms have very high resolution and permit some subject movement. One of the more famous stereograms is Lloyd Cross's "Kiss II." In this stereogram, Pam Brazier winks and blows a kiss as the observer moves left to right.

If you should try to create such a hologram, start off by determining the proper exposure you need to make a single-strip hologram. Keep one slide in the setup, and make 5 to 10 exposures on the plate, making sure to move the plate with each exposure. Develop and choose the best exposure time. Use this exposure time to create the stereogram.

Using an LCD display

You can eliminate the slide, lens, and screen from the stereogram setup and replace it with an LCD screen, such as the inexpensive black-and-white LCD TVs from Radio-Shack. By doing so, you can open up a whole new field of research and technology. There are two types of LCD screens on the market: *transmissive* and *reflective*. Obtain the transmissive type.

By using an LCD screen, you can shoot standard videotape and play it back frame by frame through a suitable VCR to make a stereogram. Or, you can create 3-D objects on a computer and rotate it frame by frame on the LCD screen to synthesize the hologram. By using a modified LCD TV, both options are open to you. The LCD TV has a radio frequency (RF) input that can accept signals from VCRs and the RF output from computers. If you use a computer, you can also program it to control the shutter and move the film plate with a linear stepper motor. (This only applies to anyone who is capable of interfacing the computer to carry out such tasks.) The computer need not be an expensive or high-powered model. I used a Commodore 64 computer. The resolution

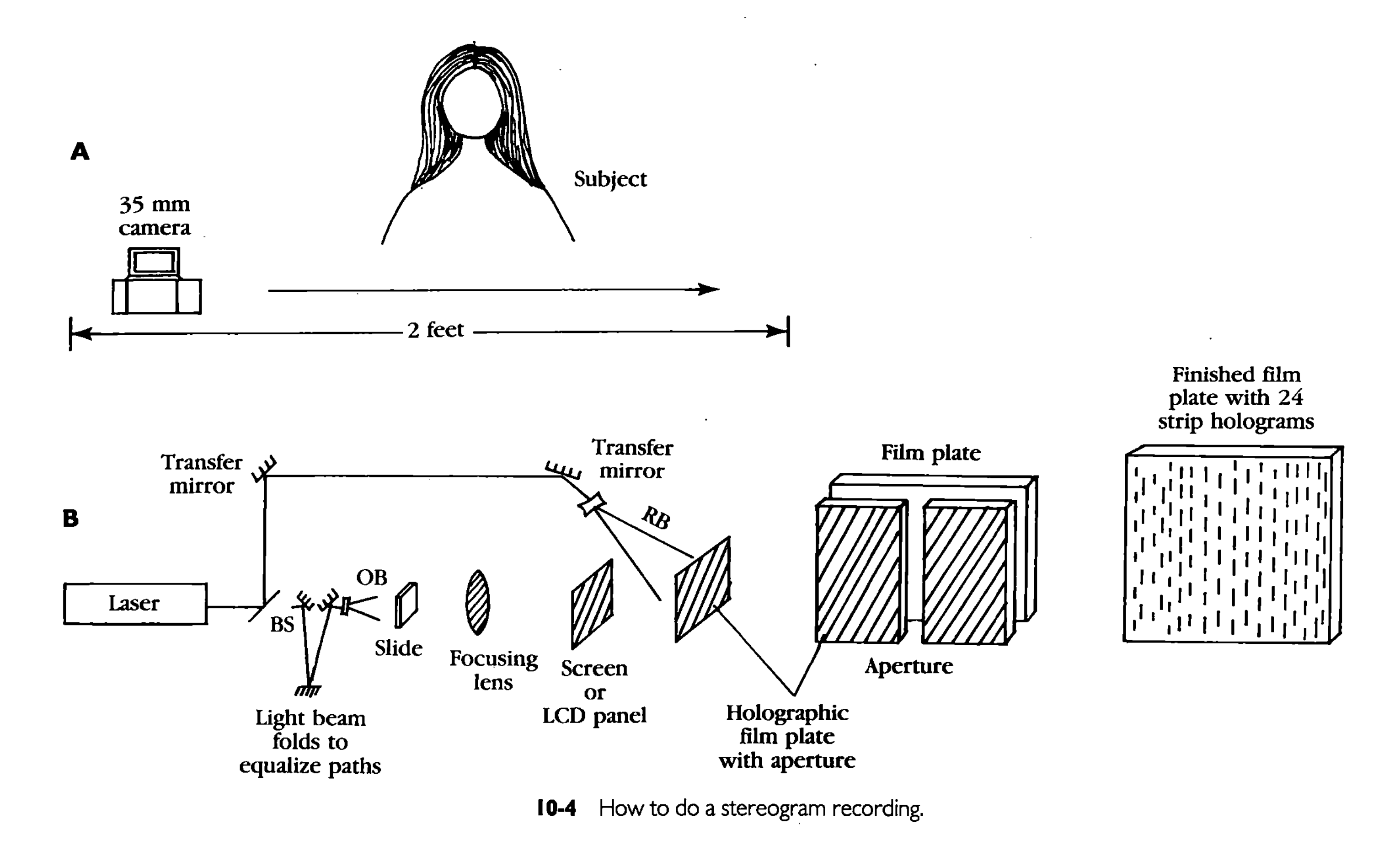

10-4 How to do a stereogram recording.

from this computer isn't very high, but neither is the resolution of the inexpensive LCD TV I used. Today there are high-resolution color LCD screens with 700 × 400 pixel resolution that are EGA and VGA compatible. Some screens are used in connection with overhead projectors; the contrast and resolution of these units are far superior to anything I have worked with. (I have a patent pending for using an LCD screen to generate holograms.)

In addition to making standard stereograms, these LCD panels can be used to create various masks for pseudo-color and natural-color holography. To use an LCD panel or TV, place the screen at the location of the paper screen. The focusing lens and slide are no longer needed. Step through the images on the LCD screen, making an exposure and moving the film plate for each step.

8E75 ACETATE FILM

Thus far, concentration has been on using glass film plates to simplify things for beginners. But when you have considerable "table time" under your belt, you might begin to wonder what you're missing by not using film. The biggest advantage film has over glass plates is cost. A 4×5-inch glass plate costs about $5. A 4×5-inch sheet of film costs about $1.50. In addition, although both the film and plates have the same type of emulsion material, the film is usually 2 to 3 times faster than the plate, so you can reduce your exposure time to about one-half of what you are using with plates. For those of you working with a low-powered HeNe laser, this is a great advantage.

Using film

The easiest way to use the film is to sandwich it between two glass sheets. Sometimes when shooting a glass plate, you only need one plate holder to help eliminate shadows and provide a little more working room. You cannot do this with a film sandwich. You must use two plate holders, one on each side, to prevent the film from moving between the glass. If the film moves even slightly, no hologram will be recorded. When using film, don't forget to check for the emulsion side and have it facing towards the subject.

The disadvantage to the glass sandwich setup is that internal reflections, caused by the glass, record onto the film. If you are using a polarized laser, this helps somewhat to prevent these reflections. But the only way to really get rid of them is to use an index-matching fluid between the glass and the film. When taking a holography class offered by Linda Law, she had a good type of index-matching fluid: clear (not colored) lamp oil.

The procedure is as follows. Get a piece of glass slightly larger than the size film you intend to use. Make sure the glass is perfectly clean before you start. Do all following procedures under safelight illumination. On a level table, pour a little lamp oil into a small puddle in the center of the glass. Remove a sheet of film and check for the emulsion side. Place the sheet of film emulsion side up onto the glass plate. Try not to get any oil on the emulsion side. Place paper towels on top of the film, making sure to cover all the edges of the film. The paper towels absorb any excess oil that will be pressed out in the next step. Using a rubber roller and with firm pressure, roll over the paper towels, pressing the film to the plate to remove excess oil and squeeze out any air bubbles. Remove the towels and look at the plate under the safelight. If you can see any air bubbles, try to press them out with the roller, being careful the roller is not oily. These air bubbles will show up as imperfections or black dots on the finished hologram, so try to remove as many as possible. If necessary, start over from the beginning. You might try bowing the film at the center when placing it on the glass plate to provide a more even displacement of oil as the film falls into place.

Once you have finished getting rid of most of the air bubbles, you can use the plate. Notice that the film sticks to the glass very firmly, so you do not need to use another piece of glass in front. Pick up the film/glass and place it in a plate holder with the emulsion facing the object you are holographing.

RED-SENSITIVE DCG

Jeff Blyth is a U.K.-based holographer. He has researched DCG emulsions and developed a formula for red-sensitive DCG. If you are interested in pursuing this material to make holograms, refer to the winter 1989 issue of *Holographics International* magazine; the address and phone number are listed in the "Sources" section. In this issue, Jeff Blyth outlines the preparation of the formulas and how to apply the compound to glass plates for exposure. Although the formulas and procedures do not appear very difficult to accomplish, the main component of the compound is volatile and caustic in pure form, so follow the precautions as described in the article. After the compound is neutralized and mixed with gelatin, it becomes reasonably safe. In addition, the pH of the stock solutions must be carefully monitored during preparation to ensure you produce workable solutions. If a standard pH meter isn't available to you, Jeff recommends using a portable pH meter designed for use with swimming pools.

PULSE LASER HOLOGRAPHY

When I began this book, I wasn't aware of the availability of any surplus ruby rods on the market (new rods cost far too much for me to have considered them a viable option). I have since located a source for red ruby rods suitable for pulse holography, but unfortunately time and space restraints limit the discussion to a cursory overview and mention of the sources I have located.

The advantage of pulse holography is the ability to shoot living subjects, such as people and flowers, because a pulse laser system supplies all the necessary energy to expose a film plate in a millisecond. The problem of vibration and movement is greatly reduced (no table is required). However, with the simple system outlined shortly, holographing people's faces to produce a portrait is out of the question, but you might want to try holographing somebody's hand holding an object or flowers.

You must be very careful with pulse systems. The high charge on the capacitor bank required to fire the flashtube is lethal. Secondly, the laser beam pulse from the rod can cause a sheet of paper placed in its path to give an audible snap—look at the paper where the laser beam hit, and you'll see ablation (burned spot). *Never look into the barrel of a pulse laser system. An accidental firing will cause irreparable optical damage.*

I purchased a ruby rod 1⁄10 inch (2.5 mm) in diameter by 3½ inches long with mirrors for $200. The diameter of the rod might seem small, and it is; rods are typically ¼ inch diameter. But if you compare the 1⁄10-inch (2.5 mm) diameter to the typical diameter of an HeNe laser, which is 0.5 to 1.5 mm, the diameter of the small rod is in most cases twice as large. This rod requires 200 to 400 joules of energy to fire. Flash lamps used in photography are sometimes rated in watt/seconds, which is equal to joules. So a flashtube rated at 100 watt/seconds is 100 joules. A typical flash unit used on a 35 mm camera is about 7 watt/seconds. Professional flash units do attain the necessary power required to fire laser rods, but that would be a prohibitively expensive way to go. You can put together a suitable flash system for about $50. Flash lamps and the electronics are available from Allegro Electronics and Dick Anderson.

The rod I purchased has its mirrors attached to the rod. The rod operates in the TEM_{00} mode (a requirement). The coherence length of the rod is short, only about one-half angstrom (0.05 nm), so I will be starting by shooting relatively flat holograms. Because the mirrors are attached to the rod, I cannot at this time use an etalon or chemical Q-switch (explained in chapter 12). But depending on the degree of success I achieve, I can opt to remove one of the mirrors from the rod and try to put together an etalon out of optical flats to increase the coherence length of the laser rod. This would lead to putting a Q-switch together to recover the power loss from using the etalon. Regardless of these possible manipulations, there are some basic construction principles you should be aware of.

Construction notes

Treat the rod carefully. Whenever you plan to handle it, wear white cotton gloves to prevent any oil, dirt, or fingerprints from your hands from getting on the rod. If you should get any dirt on it carefully clean it off using rubbing alcohol. To mount the rod, I advise buying a glass tube. The glass tube should have an inside diameter that closely matches the diameter of your ruby rod. Heat the glass and pinch one end of it to prevent the rod from falling through. When the glass cools, insert the ruby rod into the glass tube. On the open end, use a small amount of epoxy to secure a piece of optically clear flat glass to the rod.

The glass tube serves two purposes. First, it makes handling the rod easier because you never really touch the rod, just the glass tube. Second, the glass tube acts as a UV block to help prevent solarization of the rod. Flash lamps typically give off a lot of UV light when fired. Even though the glass quartz tube used in the flash lamp construction does stop some of it from escaping, the glass tubing around the rod blocks even more UV.

Solarization of the rod is caused by UV light. It begins as a frosting or white filmy buildup on the walls of the rod. It continues with each firing of the rod. Eventually, more and more light energy is required for the rod to reach its threshold and fire in order to compensate for solarization. The only cure for this is to have the diameter of the rod cut down on a lathe to remove the white coating, a costly operation. It's best to prevent solarization from happening in the first place.

Once you have your flash lamp and glass-mounted rod, mount both components in a mirrored container that focuses as much light onto the rod as possible when the flash tube fires. Many systems use an elliptical mirror.

When you are ready to use the system, I advise starting with simple single-beam setups. Some professional pulse systems have a modeling light or laser beam that is used to align all the optical components. With this "cheap-and-dirty" system, we don't have that luxury. You must fire the laser to check its beam path, so it's to your advantage to use a single lens that spreads the beam for single-beam holography. Before firing the laser, make sure there isn't anything around the laser that could reflect the beam into your eyes.

FURTHER READING

Benton, S. A. "Achromatic Images from White Light Transmission Holograms." *J. Opt. Soc. Am.* 88 (Oct. 1978): 1441A.

Benton, S. A., H. S. Mingace, Jr., and W. R. Walter. "One-Step White Light Transmission Holography." Proceedings of the SPIE, 212 (1979): 2–7.

DeBitettro, D. J. "Holographic Panoramic Stereograms Synthesized from White Light Recordings." *Applied Optics* 8 (1969) 1740–1741.

Molteni, W. J. Jr. "Black and White Stereograms." Proceedings International Symposium on Display Holography (1982).

Smothers, W. K., B. M. Monroe, A. M. Weber, and D. E. Keys. "Photopolymers for Holography." Practical Holography IV, SPIE (1990) (in press).

Tamura, P. N. "Multicolor Image from Superposition of Rainbow Holograms." SPIE 12, (1977): 59–66.

Weber, A. M., W. K. Smothers, T. J. Trout, and D. J. Mickish. "Hologram recording in Du Pont's new photopolymer materials." Practical Holography IV, SPIE (1990) (in press).

Chapter 11

Light

Light has proven to be a mystery. Its intricacies and nature have confounded the greatest minds in science. Even today, the best theories of light are a compromise between two seemingly opposing descriptions. One view is of light when it behaves as a particle, and the other view describes the behavior of light as a wave. Both descriptions are considered correct, and physicists have named this self-contradictory theory the *wave-particle duality.* The description used depends on the given set of conditions. The best view to use in any situation is the one that explains the most with the fewest assumptions (in the simplest manner).

WHAT IS LIGHT?

Despite the difficulties inherent with the wave-particle duality, it is possible to define the properties of light. First, light is a form of radiant energy, a small section of energy that is an element of the electromagnetic spectrum. What at first appears to be unrelated energy such as radio waves, microwaves, and x-rays are in fact also part of the electromagnetic spectrum. Electromagnetic energy has a number of common characteristics. One such characteristic (primarily dealt with in this book) is that they propagate in waveform. Electromagnetic waves can be easily visualized. They look like waves that are created when a pebble is tossed into a pool of still water (see FIG. 11-1). The water waves are like electromagnetic waves that radiate in every direction from their point of origin.

Figure 11-2 is a diagram of the electromagnetic spectrum. Light differs from radio waves or x-rays in frequency. Frequency is inversely proportion to electromagnetic wavelength.

ELECTROMAGNETIC ENERGY

In 1820, Hans Christian Oersted discovered that electric currents generate magnetic fields (FIG. 11-3). This led to the question whether magnetic fields in turn generate electric fields. Two researchers, Michael Faraday (1791–1867) in England and Joseph Henry (1797–1878) in the United States, independently answered the question affirmatively. Figures 11-4 and 11-5 illustrates two experiments that demonstrate electric field generation from magnetic fields.

The link between electricity and magnetism is clearly established by these

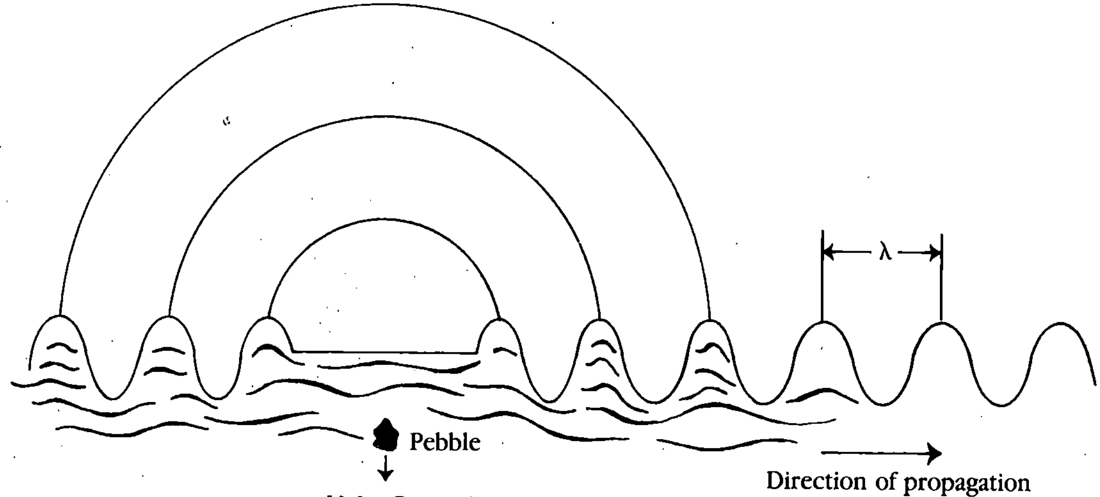

11-1 Generation of waves by a falling stone.

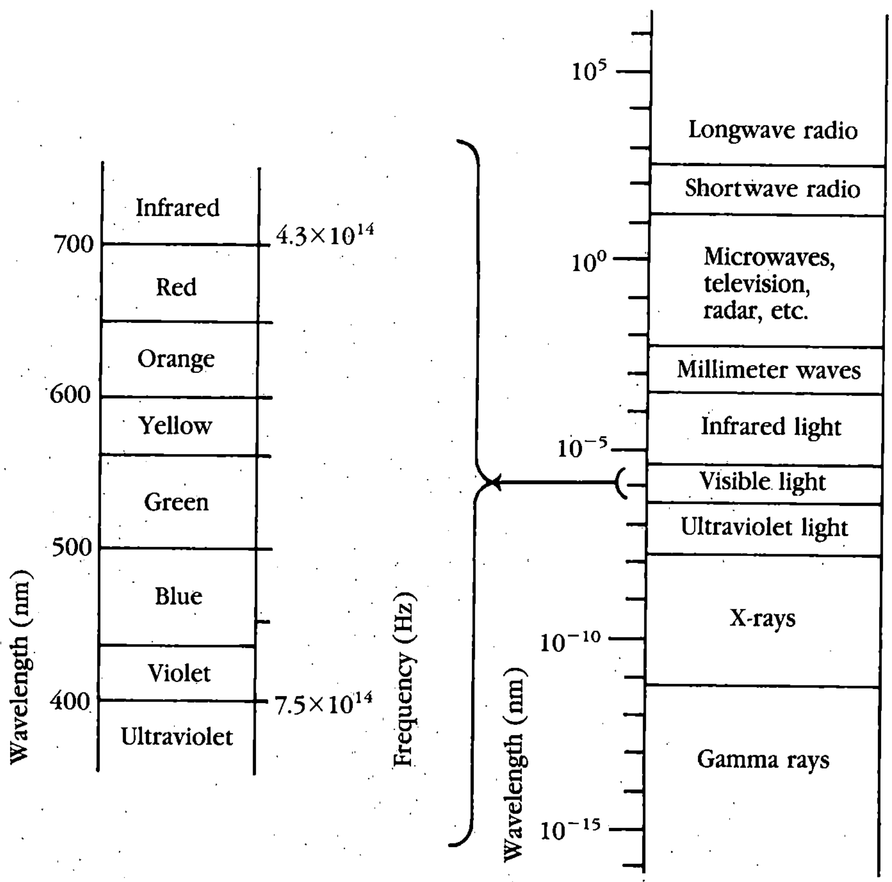

11-2 Electromagnetic spectrum. White is the equal combination of all colors, and black is the total absorption of all colors.

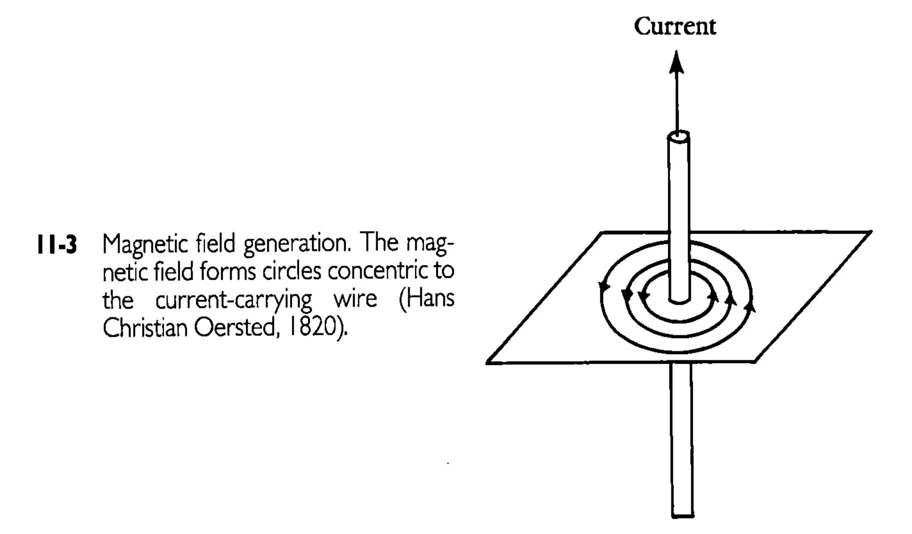

11-3 Magnetic field generation. The magnetic field forms circles concentric to the current-carrying wire (Hans Christian Oersted, 1820).

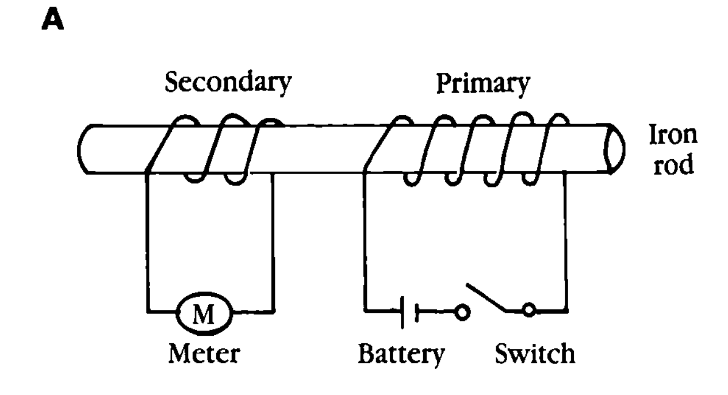

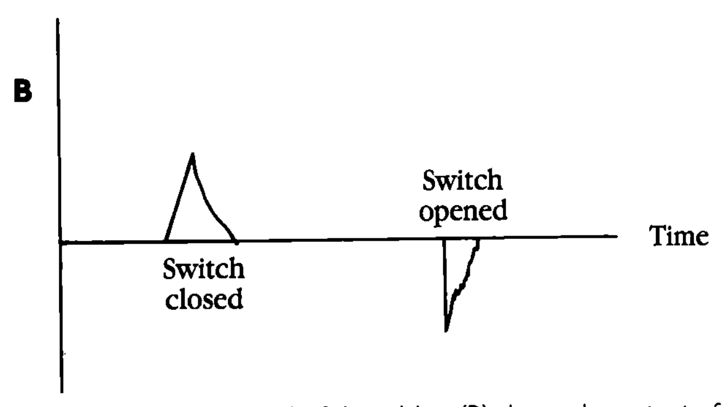

11-4 Induction (A) show a pictorial of the wiring; (B) shows the output of the meter current in the secondary with the switch closed and opened.

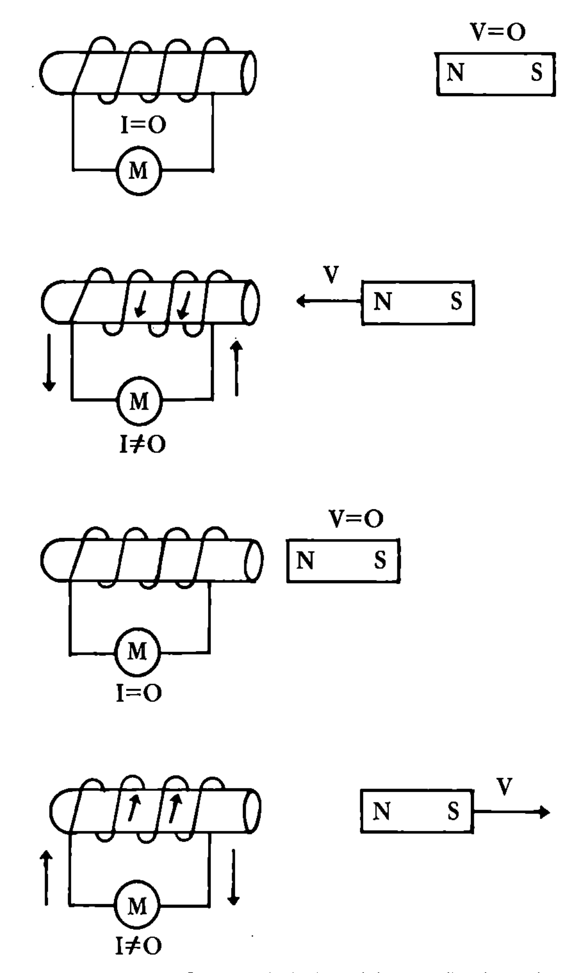

11-5 Electric field generation. Current is induced in a coil when the magnetic flux through it is changing.

experiments, but how does this relate to light?.Scottish physicist James Clerk Maxwell in 1865 recognized that electricity and magnetism could be summarized in terms of magnetic and electric fields. He discovered that oscillating electric charges radiate electric and magnetic fields. Further, the speed these fields move through space is the same as the speed of light. Maxwell didn't believe this to be coincidence. He made a bold supposition that he had discovered the nature of light waves to be electromagnetic. He was correct. Maxwell's theory was first confirmed in 1887 by Heinrich Hertz. Hertz was the first to consciously generate and detect electromagnetic waves using an LC circuit. Subsequent experiments also confirmed Maxwell's theory.

Generation of electromagnetic wave

Figure 11-6 details generating an electromagnetic wave. The two rods in the diagram represent a half-wave antenna. The generator produces a sinusoidal

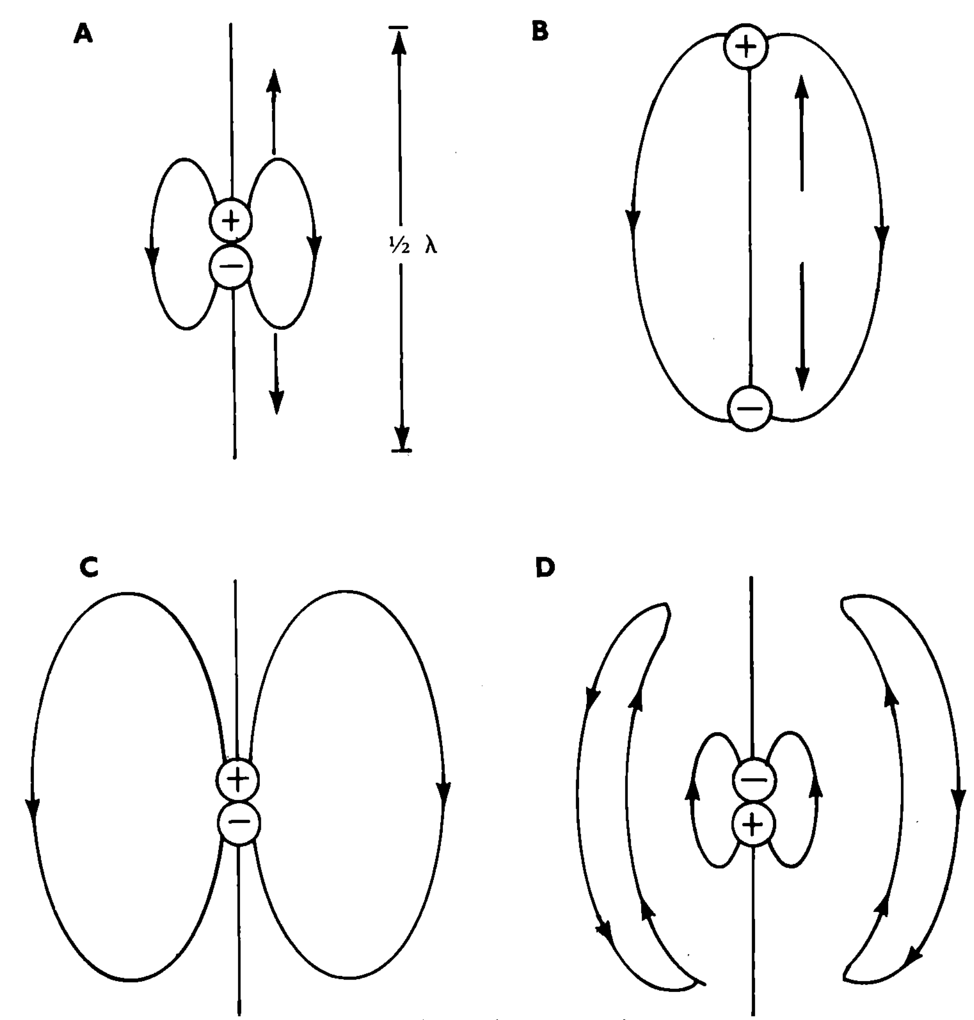

11-6 Propagation of electromagnetic waves.

wave that is represented as the dipole charges on the antenna. In A, when the oscillator switches on, the charges move in a sinusoidal manner. The positive charge moves upward while the negative charge travels downward. While the charges move, current flow is maximum. The lines of force shown are the electric fields. The magnetic fields are also created simultaneously and form concentric circles around the charges. Although the magnetic field is not shown in this diagram, you can get an idea of what they would look like by looking back at FIG. 11-3. Both the electric and magnetic fields propagate outward.

In B, charges have reached the limit of movement. The electric field potential is at its peak. Since the charges are not moving at this instant, the current flow is zero. In turn, the magnetic field collapses. The magnetic field that has already been created and propagated does not disappear but continues to travel outward.

In C, the charges reverse direction. The electric field, however, is still in the same direction. Current flows in the opposite direction, causing the magnetic field to reverse and propagate outward.

In D, the charges pass through zero, closing the electric field loop. The electric field reverses direction. The magnetic field, however, remains in the same direction because magnetically a positive charge moving downward is equivalent to a negative charge moving upward.

Remember, as the charges on the antenna are moving, the electric and magnetic fields are constantly propagating outward, and once separated from the antenna, they are no longer dependent on the antenna. The oscillating charges in the antenna transmit electric and magnetic fields that constitute electromagnetic waves. You might have noticed that a half-wave antenna was used. The length of an antenna is a practical consideration; it's important that the charges alternate back and forth within the antenna in synchronous step with the frequency. When this condition is met, the antenna is frequency-matched to the oscillator and *resonance* occurs. When the oscillator and antenna are so matched, maximum wave amplitude is transmitted.

Analysis of a wave

Figure 11-7 shows a sinusoidal wave. The wavelength of the wave can be measured from any two points that depict one complete cycle. A *cycle* is indicated by a starting point on the waveform that ends when that same point is repeated. The figure shows cycles as measured from peak to peak or between zero-crossing points, but you could use any two points, and all measurements would be identical. The most common measurement used is from peak to peak. The Greek letter lambda (λ) is used to denote wavelength.

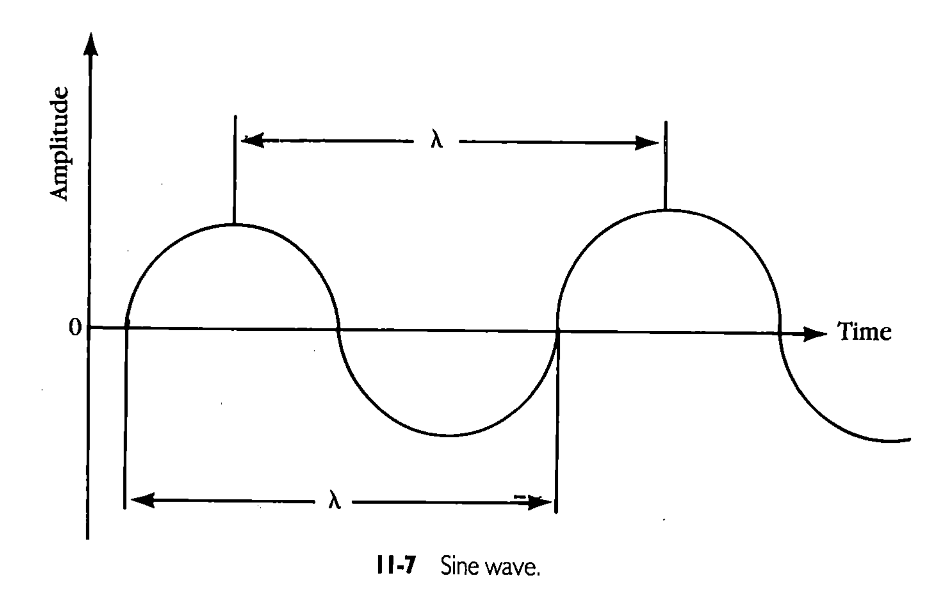

11-7 Sine wave.

Period is the measurement of time it takes for one cycle to occur. Frequency is a measurement of the number of cycles that occur in one second. Initially, frequency was written in *cycles per second* (cps). Now cps is renamed *hertz,* in honor of Heinrich Hertz, the first person to (deliberately) transmit and receive electromagnetic energy. Hertz is abbreviated Hz.

Figure 11-8 details a more accurate picture of an electromagnetic wave. It shows the component electric and magnetic field components propagating as two sinusoidal waves that are perpendicular to one another and perpendicular to the direction of travel.

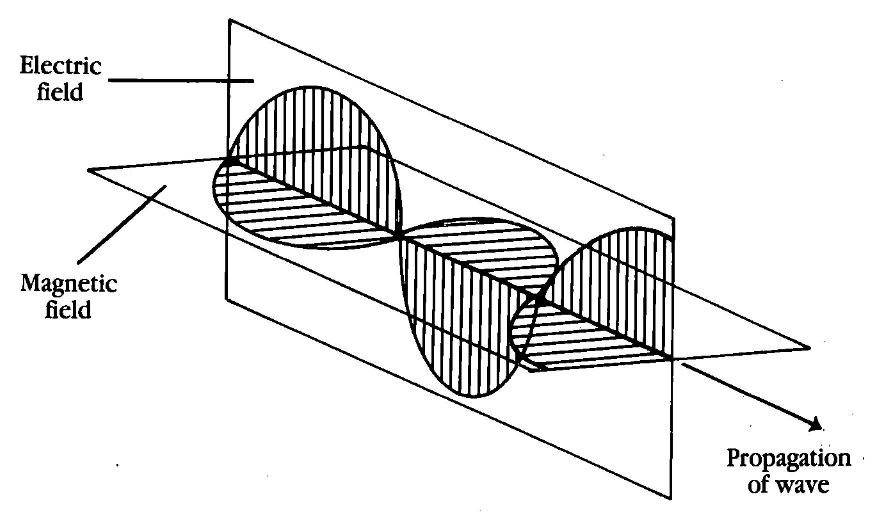

11-8 The electric and magnetic fields are at right angles to one another and to the propagation of the resulting wave.

Speed of propagation

Light, like all forms of electromagnetic energy, travels at a constant speed of 300,000,000 or 3×10^8 meters/second in a vacuum. The velocity of light decreases when traveling through a transparent material such as air or glass. The decrease in light speed due to the earth's atmosphere is small and won't affect any calculations, so it can be disregarded for the purposes of this book. The velocity at which light travels is a fundamental value in physics and is usually represented as the letter c. You might have seen the value of c given as $c = 299{,}792{,}458$ meters per second or $c = 299{,}792.458$ km/second or 186,282 miles/second. The value of light speed can be rounded off to 300,000,000 meters/second and put it in scientific notation. The value of c in this book is in meters and scientific notation as $c = 3 \times 10^8$.

Wavelength and frequency

The wavelength of light varies in proportion to its frequency. The higher the frequency, the shorter the wavelength. In holography, light is often referred to in terms of wavelength, or color. The following mathematical expressions allow you to find wavelength if you know the frequency or vice-versa.

$$\text{wavelength} = \text{light speed/frequency}$$
$$\lambda = c/v$$

$$\text{frequency} = \text{light speed/wavelength}$$
$$v = c/\lambda$$

EXAMPLE 1:

Light with a frequency of 4.73×10^{14} has a wavelength of:

$$\lambda = 3 \times 10^8/4.73 \times 10^{14}$$
$$= 6.33 \times 10^{-7}$$

EXAMPLE 2:

Light with a wavelength of 633 nm has a frequency of:

$$v = 3 \times 10^8/633 \times 10^{-9}$$
$$= 4.73 \times 10^{14} \text{ Hz}$$

GENERATING LIGHT

Although I recently discussed generating electromagnetic waves using an oscillator and antenna, it is not feasible, despite the fact that light is an electromagnetic wave, to generate light that way. The reason is that we aren't capable of building an electronic oscillator whose frequency lies in the 4×10^{14} Hz to 8×10^{14} Hz range. Secondly, we can't build an antenna small enough to operate at the tiny wavelengths of light. The question of how light is generated remains.

In the following discussion, think of light as if it were a particle called a *photon* or *quanta.* This does not imply abandonment of the waveform model of light; it is just on the back burner for the moment. To understand how light is generated, let's begin at the atomic structure of matter. Figure 11-9 shows the Rutherford model of the atom, which is probably the most familiar model of the atom to the average person. This atomic model shows the nucleus at the center of the atom with electrons orbiting around the nucleus.

The number of electrons and protons contained within the atom determine the type of material as well as its electrical and chemical properties. The orbits of the electrons are also called *levels* or *shells.* The terms *orbits, levels,* and *shells* are used interchangably in the text. The shells are arranged as shown in FIG. 11-10.

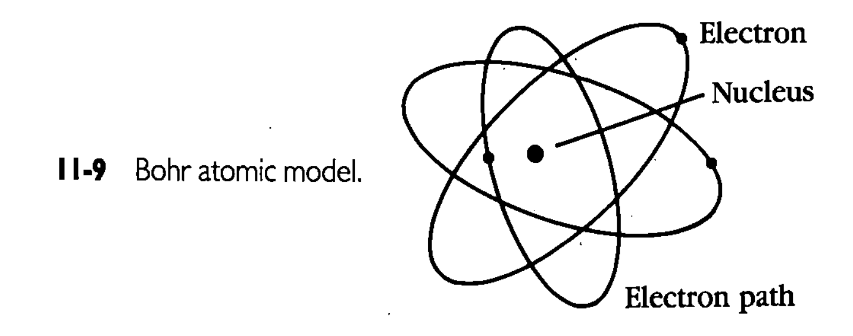

11-9 Bohr atomic model.

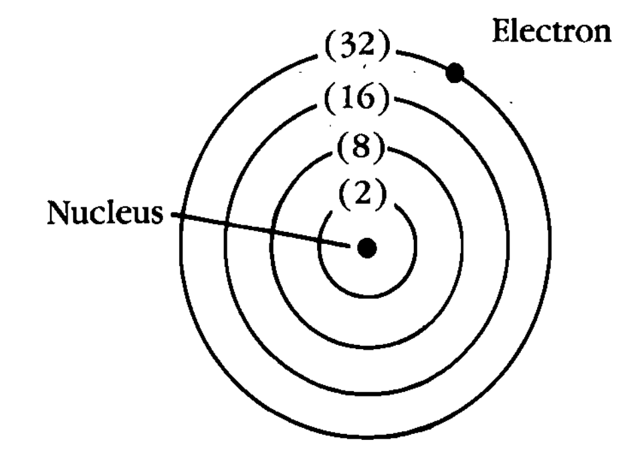

11-10 Electron orbital shells. The maximum number of electrons for each shell is shown in parentheses.

In the early 1900s, the scientific community realized that a traveling or vibrating electric charge or particle like the electron should emit electromagnetic energy. This in accordance with Maxwell's electromagnetic theory in 1865. The orbital procession of the electron circling the nucleus can be considered essentially a vibration. This being the case, the electron should emit radiation and in doing so gradually lose energy and fall into the nucleus. Obviously this is not the case. The Rutherford nuclear model, although contradictory to the electromagnetic theory, explained much experimental physics data, so scientists therefore did not want to abandon it. It became a challenge to explain the discrepancy between the atomic model and the electromagnetic theory.

To save the atomic model, Neils Bohr built upon the work of Planck in 1901 and Einstein in 1905 that had shown that energy is not always absorbed or emitted continuously but can be absorbed or emitted in discrete bundles of energy called *quanta.* This was relatively heretic at the time because it ran straight against the current beliefs.

Bohr made two assumptions. The first was that electrons could only orbit the nucleus in specific orbits; the second was that an electron in the specific orbit would not emit any radiation. For an atom to emit or absorb radiation, it did so in little packets of energy called *quanta,* whereby an electron jumps from one allowed orbit to another allowed orbit.

If an atom absorbed a quanta of energy, an orbiting electron would jump to a higher shell, becoming what is termed an *excited atom.* The jumping of an electron to different shells is called a transition. If an atom absorbs so much energy that the electron essentially leaves the atom, the atom is considered to be *ionized.* For an excited atom to emit energy, an electron jumps from a higher to a lower shell, emitting a quanta of energy. An atom with all its electrons in their lowest possible energy levels is in ground state.

The Bohr model was successful, it predicted with great accuracy the spectrum lines of hydrogen, which were well known but unexplained. This model of the atom became known as the Bohr model.

Figure 11-10 shows a hydrogen atom with the first four energy levels (shells). An atom can have many more energy levels than illustrated here. The electron depicted in the diagram is in level 4 and is considered to be excited. Atoms typically cannot remain in an excited state very long, usually about 10^{-8} second before spontaneously releasing its excess energy and emitting a photon of energy that in turn causes the electron to jump down to one of the lower energy levels. Figure 11-11 shows the possible jumps an electron can make. Any time the electron jumps from an outer shell to an inner shell, it emits an amount of energy (quanta or photon) equal to the difference in the energy of the levels jumped.

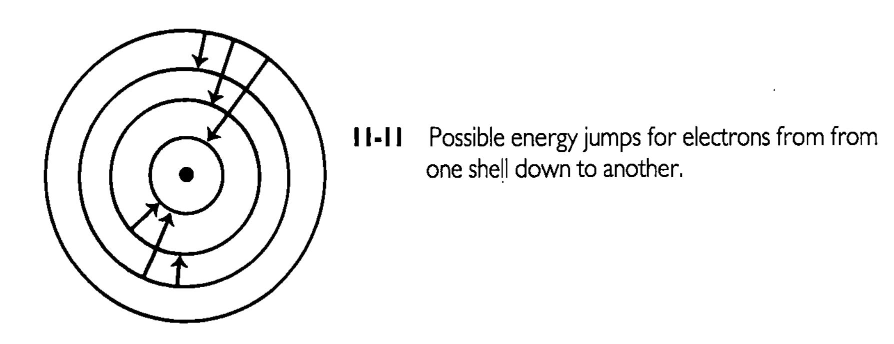

11-11 Possible energy jumps for electrons from from one shell down to another.

Figure 11-12 shows the energy levels associated with each shell or orbit of the hydrogen atom. The energy of each shell or level is given in electronvolts, abbreviated eV. (The method of determining the value of electron volts for each shell is not important for the purposes of this book.) Now suppose an electron, which is in the -0.85 eV shell, jumps to the second level, -3.4 eV. The electron emits radiation equal to the difference between levels, in this case $-0.85\text{eV} + (-3.4\text{eV}) = 2.55$ eV. Later I explain how this excess emitted energy can be used to determine the frequency and wavelength of the emitted

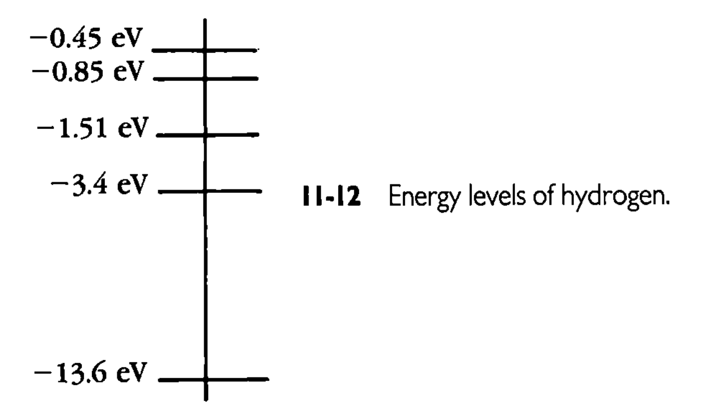

11-12 Energy levels of hydrogen.

radiation. The Bohr model has been improved and advanced since 1913 (the next advancement was the de'Broglie wave), but it isn't necessary to proceed further in the history of nuclear physics because the models shown are adequately accurate for our purposes.

Spectrum lines of hydrogen

When hydrogen emits radiation, it is always at the particular characteristic frequency determined by the electron jump to the lower energy shell. Every element has its own unique energy levels that emit specific radiation wavelengths. No two elements have the same set of energy levels. If any element is heated to the point where it emits light (becomes incandescent), it emits that light at specific frequencies. Hence, light emitted from hydrogen passes through a prism so that it breaks down into its component colors, only certain frequencies (colors) of light are present. This is the spectrum of light produced by hydrogen, which looks somewhat like FIG. 11-13A. This spectrum is called either an *emission spectrum* or a *bright line spectrum.* Some of the frequencies lie in the infrared and ultraviolet portion of the spectrum and are invisible to the eye. Each element produces its own unique spectrum.

A. Bright line spectrum (emission)

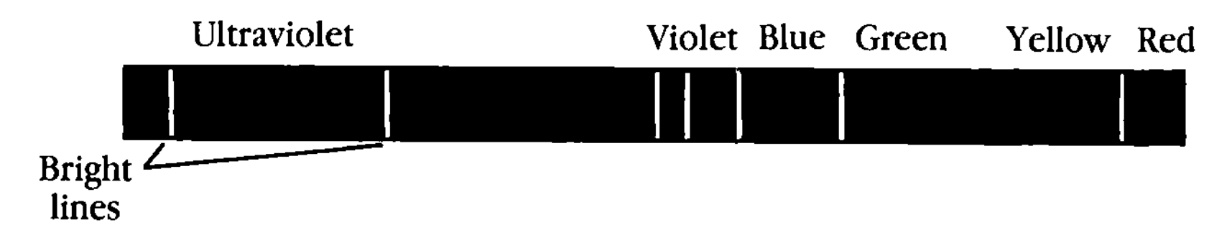

B. Dark line spectrum (absorption)

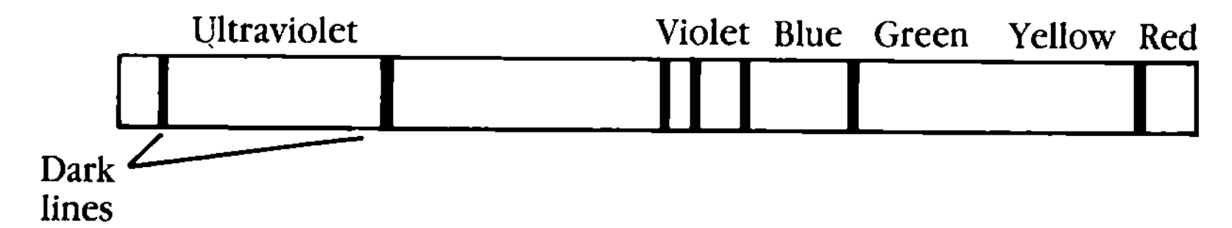

11-13 (A) Emission and (B) absorption spectrums of hydrogen. Each dark line in the absorption spectrum corresponds to a bright line in the emission spectrum.

You can calculate the wavelength of the emitted radiation if you know the the amount of energy released in the atomic transition in eV.

$$\text{Wavelength} = \frac{1242 \text{ nm}}{\text{eV}}$$

So an energy transition that generates 1.963 eV equals a wavelength of:

$$\text{Wavelength} = \frac{1242 \text{ nm}}{1.963 \text{ eV}} = 632.8 \text{ nm}$$

Dark spectrum lines An element can be heated until it emits radiation. An atom can absorb energy is through the absorption of photons. If the photon's energy matches the transitional energy required between the lower energy level and a higher energy level, it can be absorbed. If the photon has too little or too much energy, it will not be absorbed. Using this principle, if white light (which consists of all frequencies of light) passes through hydrogen gas and then through a prism, the hydrogen gas absorbs light that corresponds to the various energy transitions possible from electron jumping. Figure 11-13B shows what the spectrum would look like. There is a continuous spectrum containing all the frequencies except those absorbed by the hydrogen. This spectrum is called a *dark line* or *absorption* spectrum. The positions of the dark lines are identical to the position of the bright lines in the emission spectrum.

INCANDESCENT LIGHT

The incandescent lamp was invented in the 1870s by Thomas Edison. *Incandescence* is light emission of a material via heating it. A common house light is a tungsten filament in a evacuated bulb. The bulb is evacuated to prevent oxidation of the filament when it is heated to high temperature. In order to provide resistance to the electric current and generate heat, the tungsten is made into a long thin wire. When electricity is supplied to the lamp, the current quickly heats the filament to 2200 to 3200 degrees centigrade. The heating is caused by the collisions of the electrons from the electrical supply with the tungsten. The tungsten atoms emit frequencies from infrared through visible light, a broad spectrum of light.

You might be wondering how tungsten emits a broad spectrum light; shouldn't it produce a bright line spectrum characteristic of the element tungsten? In a metal, the atoms are much closer together than in a low-pressure gas. Because of their closeness, the outer electron shells mix, allowing electrons to jump from one atom to another. This creates a condition where virtually any energy transition becomes possible. The outcome of this is a continuous spectrum that contains all the frequencies of light. So if a bright line spectrum is produced by a low-pressure gas (such as hydrogen), increasing gas pressure would result in a broadening of the spectrum lines. Increasing the gas pressure even more would result in a broad spectrum of light.

GAS-DISCHARGE TUBES

Fluorescent lamps are another common lamp used in homes. These lamps are usually tubular or circular in design. The tube is evacuated of air, which is replaced with mercury vapor. The inside of the tube is coated with a fluorescent material.

Each end of the tube has a heater coil and cathode assembly. When current is applied, the heater coil heats the cathode, which emits electrons and ionizes the heated gas around the cathode. A ballast transformer provides a high-voltage pulse that causes an electric arc to flash from one end of the tube to the other. This electrical arc ionizes the gas throughout the tube, providing a low-current path for electricity to flow. The current flow keeps the gas in the tube ionized. The individual atoms of gas absorb energy from the current flow and are constantly jumping to excited, then to stable states, emitting ultraviolet light. This ultraviolet light is converted to visible light by the fluorescent coating on the inside of the tube.

Fluorescent materials have an interesting property of being able to absorb high-frequency light and then emit light of a lower frequency. To illustrate, look again at the Bohr model in FIG. 11-11 that shows the possible energy jumps. Imagine this diagram now represents the fluorescent atom. When the atom absorbs an ultraviolet photon, an electron jumps up to the fourth level. When the electron falls back down, it can fall into any of the lower shells. The energy it emits falling to an intermediate shell is less than it is if it falls all the way to the ground state. The frequency associated with a drop to an intermediate level is less than the frequency of the ultraviolet light and lies in the visible spectrum.

Fluorescent lamps have mercury vapor excited by the electrical current, emitting ultraviolet light. On the other hand, *phosphors* are excited by ultraviolet light and emit light in the visible spectrum.

The neon lamp is a close relative to the HeNe laser (explained further in chapter 12). The neon lamp is another discharge lamp seen mostly as store signs. Some of its operating principals are similar to fluorescent lamps. With neon lamps, there are no heater coils or phosphors. A continuous high voltage is applied to the electrodes at each end of the tube. A current-limiting resistor is placed in series with the tube to prevent overdriving the tube once the gas becomes ionized by the high voltage. The neon gas emits light in the red to orange range, so phosphors aren't required and a clear glass tube is used. The color of the light can be changed to blue by the introduction or a small amount of mercury into the neon gas.

Other types of gas-discharge lamps are available such as mercury vapor, sodium, carbon dioxide and metal halide.

SEMICONDUCTOR LIGHT SOURCES

Semiconductor light sources, I feel, represent an important future prospect to holography. Chapter 12 describes in greater detail the operation of semiconductors and their construction. Semiconductor lasers, I believe, will eventually replace gas lasers and become the laser of choice among holographers. If you become confused by the short description that follows, you might want to read the section on semiconductor lasers in chapter 12 first.

A semiconductor diode can be made to emit light. This is accomplished at the pn junction of a forward-biased diode. The semiconductor materials p and n produce hole-electron recombination at the junction where electrons jump from the n material to the p material. The energy of the jump, or the energy required to jump is analogous to the electron jump to a lower shell when an excited atom emits radiation. This phenomenon is called *electroluminescence* or *injection luminescence.* Table 11-1 lists some materials that are used to make diodes with their respective wavelengths.

LIQUID CRYSTAL DISPLAYS

Liquid crystal displays (LCDs) don't emit light but rather control it. The control of light is achieved by the structure of the LCD panel. The LCD is constructed of

Table 11-1 Semiconductor Diode Materials

Material	*Wavelength (nm)*	*Light*
Gallium antimonide	1770	Infrared
Indium phosphide	985	Infrared
Gallium arsenide	898	Infrared
Gallium arsenide phos.	650	Red
Gallium phosphide	565	Green
Gallium nitride	400	Violet

two outer plates that have a transparent electrical coating on it. These plates sandwich a layer of liquid crystals. A voltage applied between the plates changes the direction of polarization of the liquid crystals. The effect this produces is that light is either reflected from the crystals or passes through. Small selected areas of the crystals can be turned "on" or "off." These small areas are analogous to pixels and can be used to generate screen images in LCD televisions or LCD computer screens. This capability has an emerging impact on the generation of stereograms. (I started patent procedures on a method of using LCDs to generate holograms in early 1988; hopefully by the time this book is published, my patent will be approved.)

WAVEFRONTS

Although it's best to look at light as both particles and electromagnetic waveforms to understand the many properties of light, many optical phenomena are explained on the basis of the simple nature of the lightwave without any direct reference to its dual particle/electromagnetic characteristics. With this in mind, let's consider Huygen's Principle, diffraction interference, Newton's rings, and polarization.

HUYGENS' PRINCIPLE

In the mid 1600s, Robert Hooke, an English physicist, proposed that light might be a wave. This theory was improved by Christian Huygens, who theorized that light leaving a source could be considered a series of spherical wavefronts (see FIG. 11-14). Any point on the approaching wavefront can be a source of secondary wavelets.

At a significant distance from the source, the curvature of the wavefronts is so small that they can be considered to be a series of plane waves. For example, the distance from the earth to the sun is so great that we can consider the light we receive from the sun as a series of plane waves.

DIFFRACTION

The phenomenon of diffraction is simply light bending around edges of objects to a small but appreciable extent. Figure 11-15A shows light waves incident on a hole in a barrier. The light waves spread out from the hole as if the light originated from the hole. Obstacles placed in the path of the light waves cannot be expected to cast sharp-edged shadows. Experiments with shadows

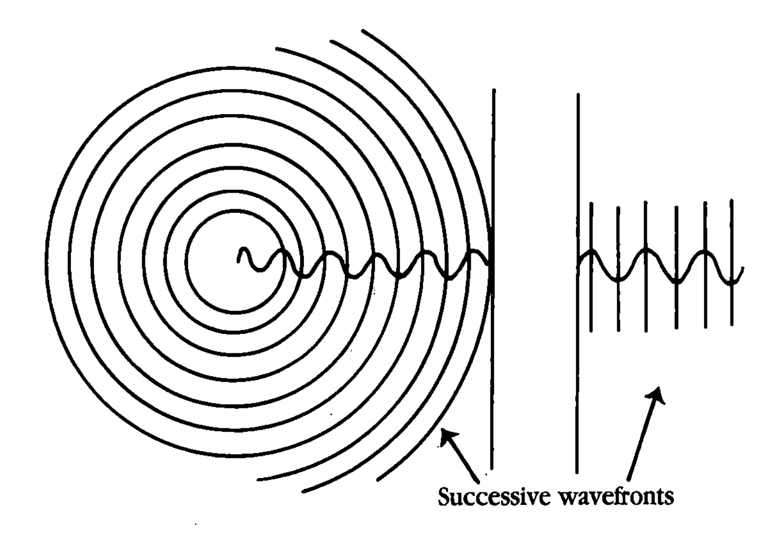

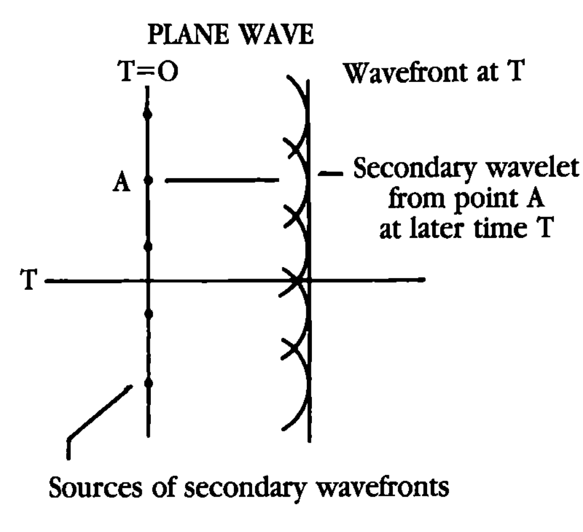

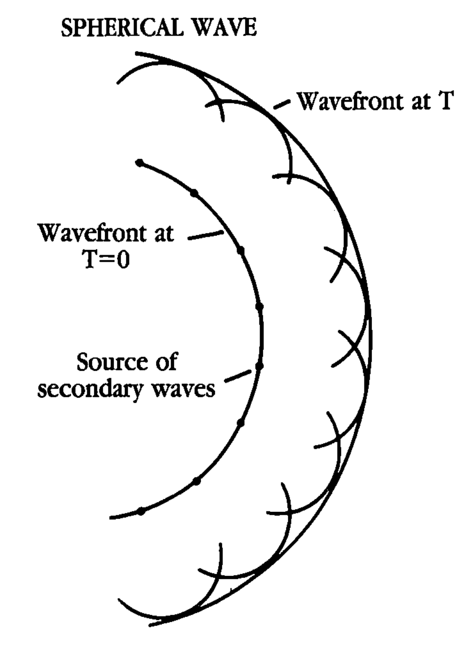

11-14 Generation of secondary waves.

as illustrated in FIG. 11-15B show the edges of shadows indeed are not sharp (this isn't easy to see because of the short wavelengths of light). The experiment can be improved by using a monochromatic (single-frequency) light source and film instead of a viewing screen. Figure 11-15C shows an enlarged section of exposed film in which the shadow edge of a razor blade shows the light and dark fringes at the edge of the shadow using monochromatic light.

Diffraction is closely related to interference. The fringes are caused by light interfering from the adjoining region above the razor's edge with the light diffracted at the edge. Frequency (and/or wavelength) plays a part in the

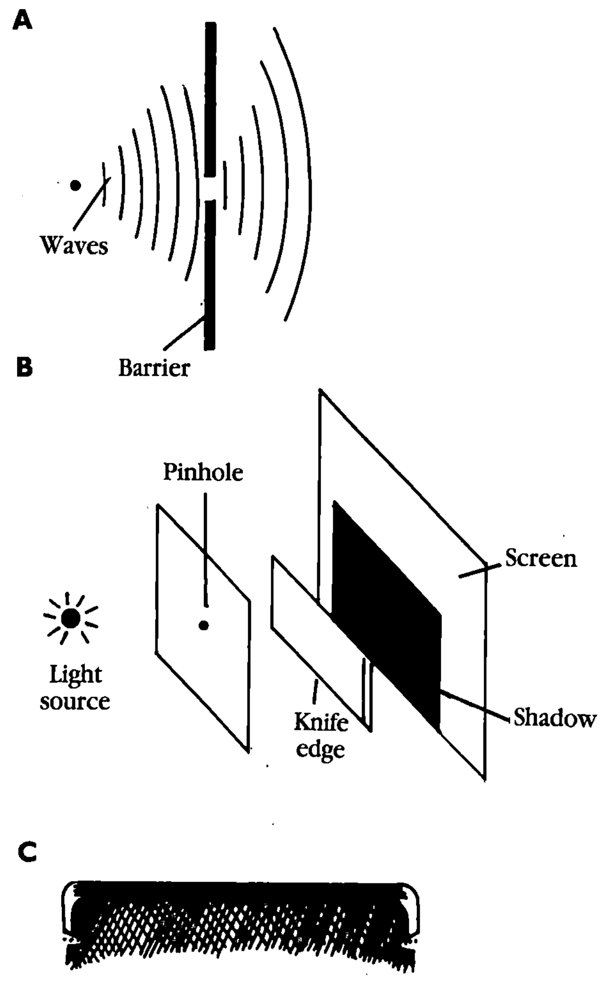

11-15 Diffraction of light. In (A), the waves spread out on the far side of the barrier as though they originate at the gap. (B) Even under ideal conditions, the shadow is never completely sharp. (C) Shadow of razor blade using monochromatic light.

amount of diffraction. Higher frequency ultraviolet light is diffracted less than red light.

INTERFERENCE OF LIGHT

The interference of light is an important concept because it's crucial in understanding the creation of a hologram. Fortunately, it's not a difficult concept. Figure 11-16 shows the basic principle of constructive interference and destructive interference. When two or more waves of the same frequency and amplitude travel past the same point at the same time, the amplitude at that point is the instantaneous sum of amplitudes of the individual waves. Constructive interference is the reinforcement of the waves in phase (in step) with one another, and destructive interference refers to the partial or complete cancel-

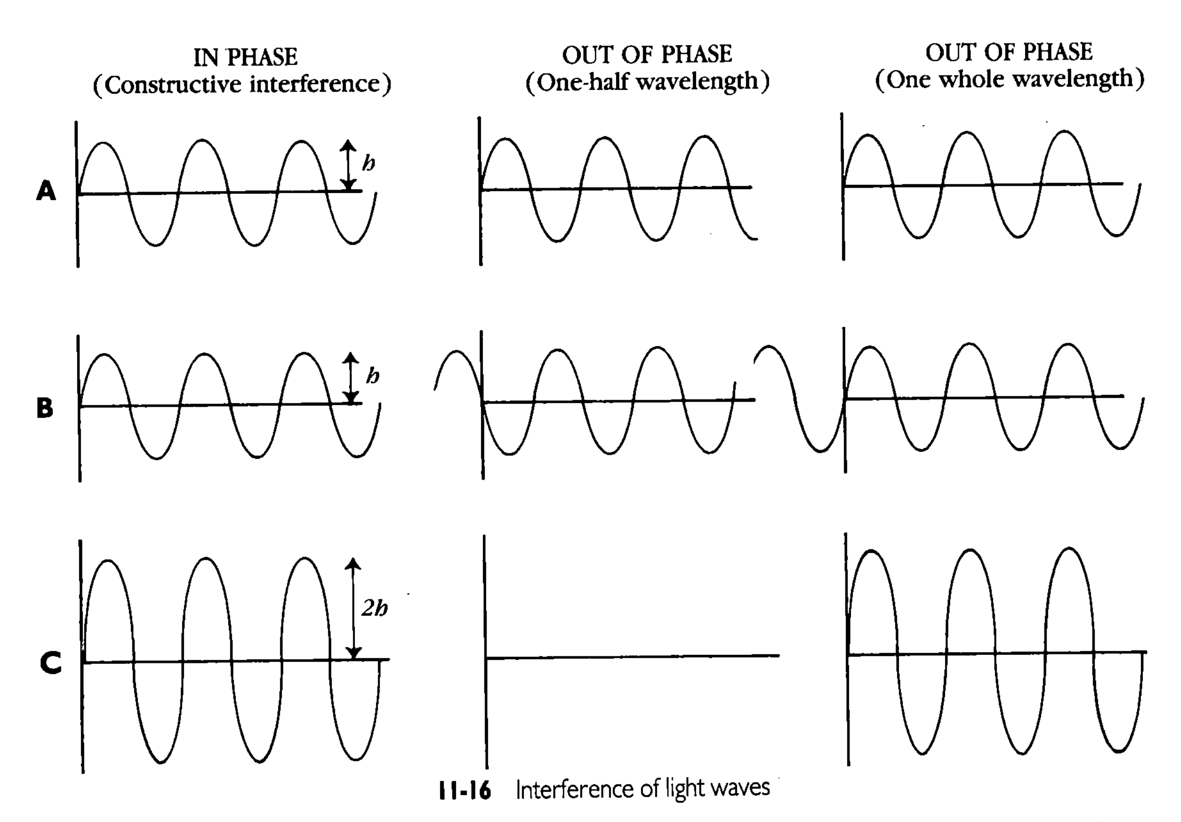

11-16 Interference of light waves

ation of waves, that are out of phase with one another. Figure 11-16B shows complete destructive interference. One wave is shifted 180 degrees out of phase with reference to the other wave. Keep in mind that total constructive or destructive interference is not always the case, there are a multitude of partial phase interferences depending upon the amount one waveform is shifted in reference to the second waveform.

Another case is FIG. 11-16C. Here a waveform is shifted one complete cycle, one wavelength, or 360 degrees, depending upon your unit of measure. The point is that again there is constructive interference. Any whole number wavelength shift like 1λ, 2λ, 3λ, etc. produces constructive interference. The algebraic sum of wave functions at a point is called *superposition*.

The interference of white light is difficult to demonstrate. Visible light is made up of a multitude of very short wavelengths. Recall that the visible part of the electromagnetic spectrum extends from 400 nm to 700 nm. Another factor that weighs heavily is the coherence of light. Light emitted from natural sources is spontaneous and randomly phased, depending on the moment the atom released the photon. The photons produced by two atoms could interfere with one another, but only if both atoms happened to emit at the same time (or very near the same time). Despite these obstacles, the interference on light was demonstrated in 1801 by Thomas Young using the double-slit experiment.

The demonstration performed by Young is illustrated in FIG. 11-17. Here a monochromatic light source is located behind a screen with a narrow slit in it with another screen in front of it with two similar slits. In front of this is placed

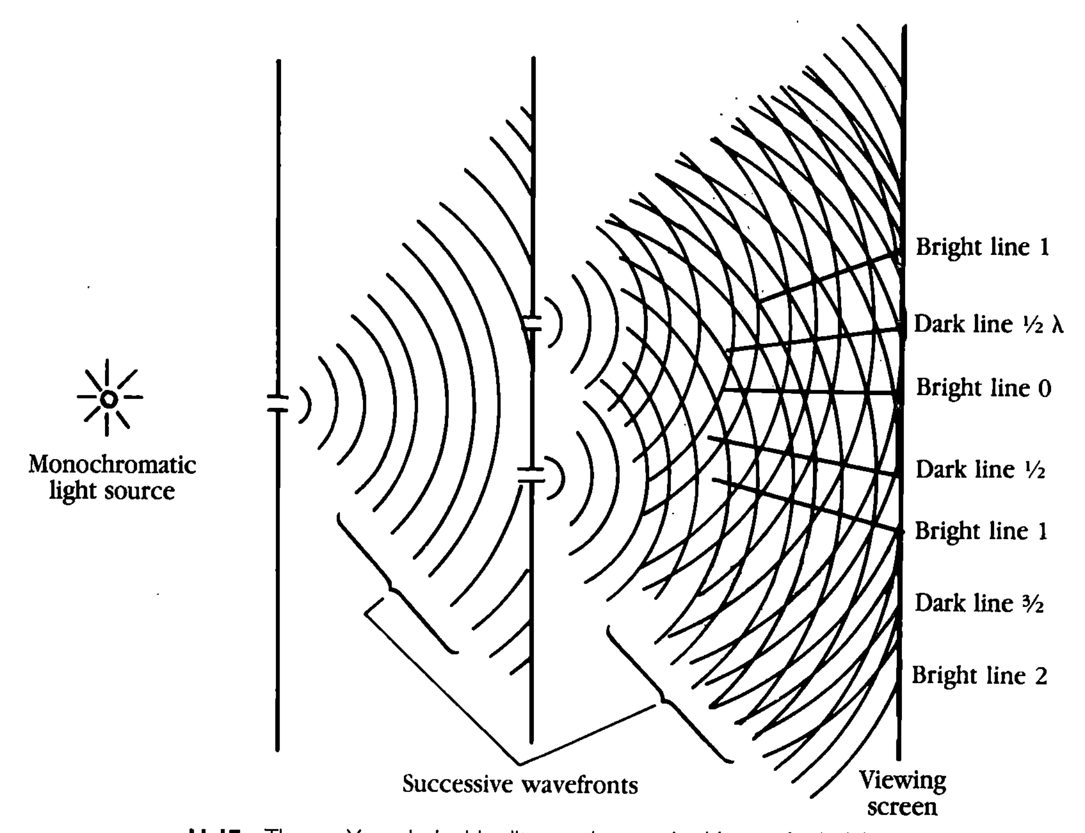

11-17 Thomas Young's double-slit experiment using Huygens' principle.

a viewing screen. The demonstration proceeds as follows. Light passing from the first split produces a wavefront in accordance with Huygens' principle. This wavefront then passes through the double slit. Each of the double slits also produce secondary wavefronts, but these wavefronts are coherent to one another because they both originated from the same wavefront propagating from the first slit. The viewing screen, owing to interference of the light waves, shows a pattern of alternating bright and dark lines.

The alternating bright and dark lines, called *fringes,* allowed Young to calculate the wavelength of light for the first time. In FIG. 11-17, the fringe labeled 0 is bright because the waves traveling to this point are in phase and reinforce each other (constructive interference). At the fringes labeled 1 and 2, the two waves again reinforce each other, producing bright lines. At these positions, the wave are a whole number of wavelengths out of phase. This is shown again in greater detail and different perspectives in FIGS. 11-18 and 11-19.

$$N\lambda = d \sin \theta$$

In this equation, N is the fringe number, d is the the slit separation, λ is the wavelength of light, and θ is the angle of deviation. Typically, N is called the *order number* of the fringe.

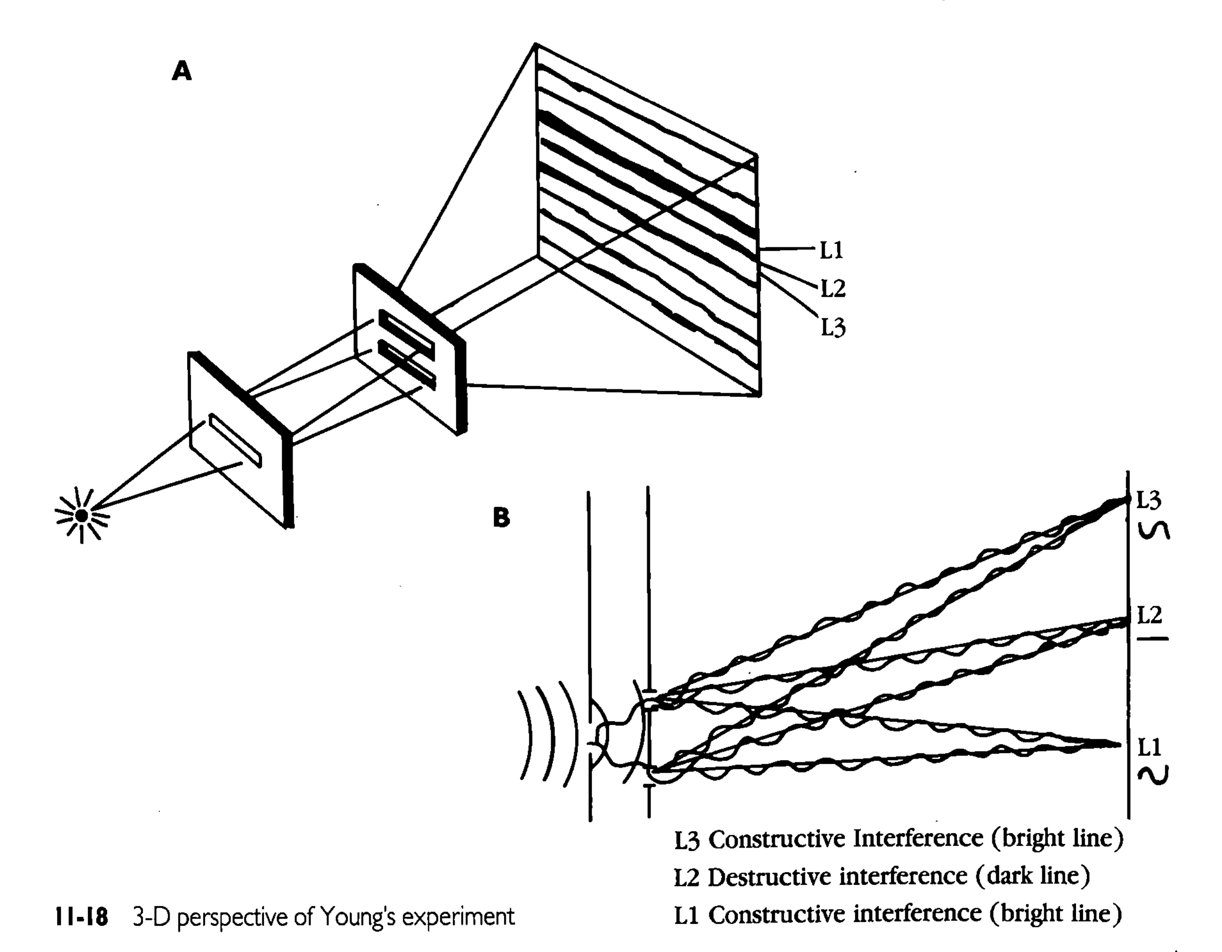

11-18 3-D perspective of Young's experiment

As stated, Young used this experiment to calculate the wavelength of light. When he performed his experiment, he used a narrow beam of sunlight because sunlight has an infinite number of wavelengths, and each wavelength has a bright fringe at a particular angle. The fringes Young observed were colored; the fringes closest to the center are blue, while the outer fringes are red.

To see how Young calculated the wavelength of light, consider the following experiment. In FIG. 11-20, the distance L is 5 meters, the slit separation is 0.00025 meter, and the distance from the center of the pattern to the approximate center of fringe 2 is 0.021 meters. Using the tan function, the angle of deviation is:

$$\tan\theta = \frac{0.021/\text{m}}{5m} = 0.0042$$

from which θ = inverse tan 0.0042 = 0.24 degrees. Young used this data to calculate the wavelength of light:

$$\lambda = d\sin\theta = 0.00025\sin 0.24 = 5.23 \times 10^{-7}$$

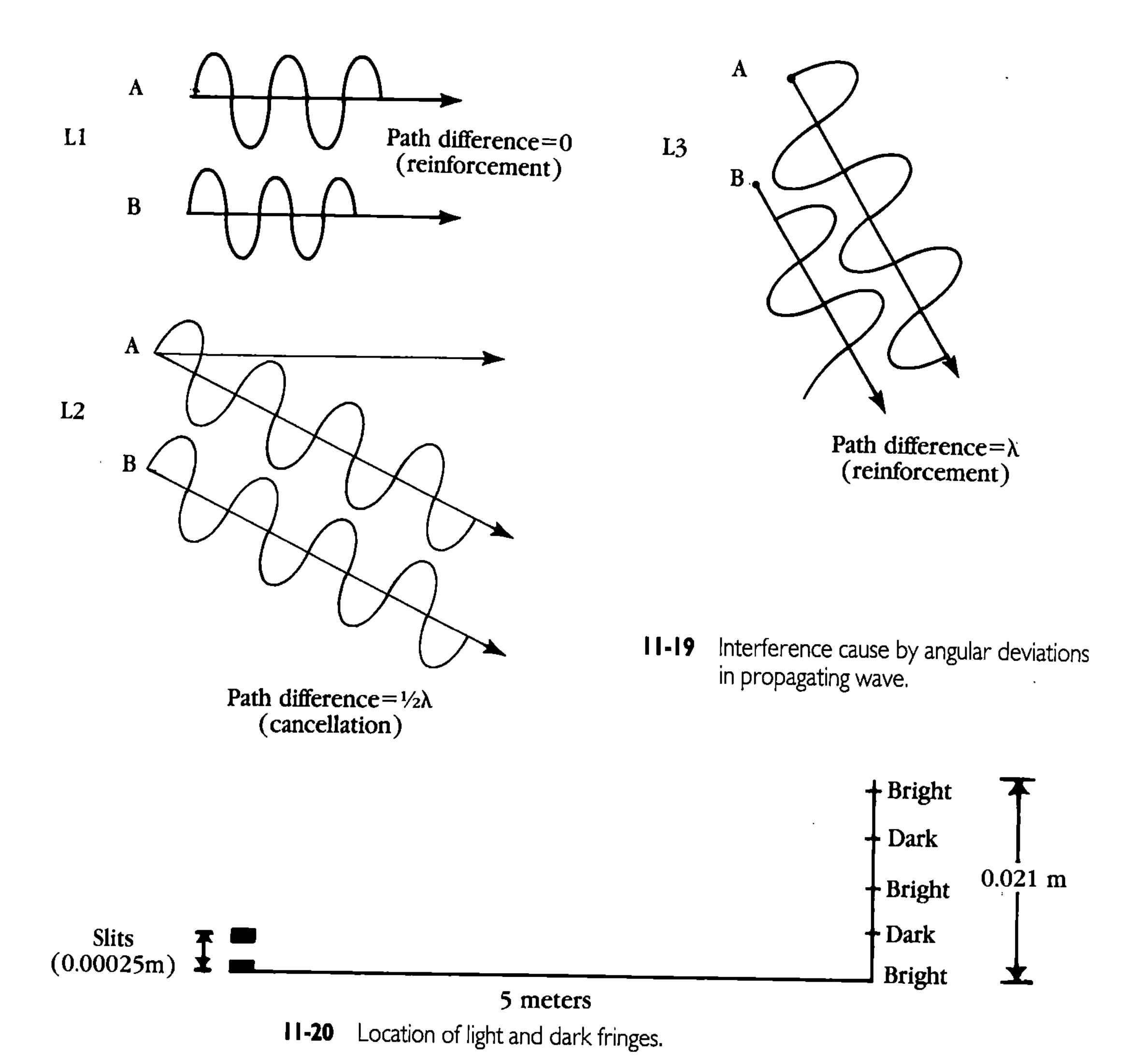

11-19 Interference cause by angular deviations in propagating wave.

11-20 Location of light and dark fringes.

Young concluded that light had an approximate wavelength of 500 nanometers, with the wavelength of blue light a little shorter and red light a little longer.

A superior monochromatic light source to sunlight is the laser. Suppose an HeNe laser illuminates a double slit with a spacing of 0.1 mm (0.0001 meter). If you put a viewing screen 5 meters away from the slits, where would the first bright line form from the center (see FIG. 11-21)? Transpose the equation

$$N\lambda = d \sin \theta$$

to:

$$\frac{\lambda}{d} = \sin \theta$$

Plugging in the numbers:

$$\frac{633 \times 10^{-9}}{0.0001} = 0.0063$$

$$\text{Inverse sin } 0.00633 = 0.36 \text{ degree}$$

Now use the tan function to determine height of fringe.

$$\tan 0.36 = \frac{\text{Opposite side}}{\text{Adjacent side}}$$

$$= \frac{\text{Opposite side}}{5}$$

Transpose to $5 \times \tan 0.36 = \text{opposite} = 0.031$ meter. The center of the first bright line from the center bright line would be 0.031 meters.

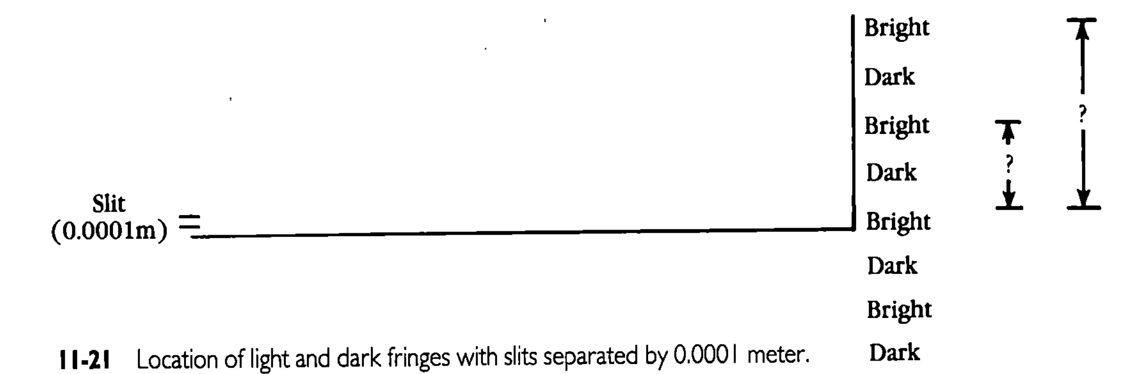

11-21 Location of light and dark fringes with slits separated by 0.0001 meter.

The second bright line would lie at

$$\frac{(2)\,(633 \times 10^{-9})}{0.0001} = 0.01266$$

$$\text{Inverse sin } 0.01266 = 0.72 \text{ degree}$$

Again, using the tan function:

$$\tan 0.72 = \frac{\text{Opposite side}}{5}$$

$$5 \times \tan 0.72 = 0.063 \text{ meter}$$

Diffraction grating The spacing of the double-slit interference pattern is different for different wavelengths. There are two difficulties in utilizing a single double-slit for analyzing light. One is that the "bright lines" are in fact faint, and secondly, the lines themselves are rather broad, making it hard to locate the centers.

A *diffraction grating* alleviates both problems. The distance between the slits become closer, which in turn spreads the line pattern. Additionally, a diffraction grating consists of a large number of parallel, evenly spaced slits. The effect of these multiple slits accentuates and defines each bright line.

Inexpensive diffraction grating is available from Edmund Scientific (see "Sources"). Edmund sells what it calls experimental grade gratings that have 600 to 1200 lines per millimeter for $12.25 each (part number J42,284). These gratings are small, the size at the price quoted is 12.5 mm × 25 mm, which is about ½ inch by 1 inch.

Less expensive than this are plastic gratings. A 8½ × 11-inch sheet sells for $7. The sheets can be cut with scissors to smaller sizes and mounted into slides or other holders. These grating sheets are quoted as having 13,400 lines per inch (Edmund part number J40,267). To compare this with the experimental grade, convert 13,400 lines per inch to lines per millimeter.

$$25.4 \text{ mm} = 1 \text{ inch}$$

$$\frac{13{,}400}{25.4} = 527 \text{ lines/mm}$$

To gain an appreciation of gratings, calculate the impact the least expensive plastic diffraction grating would have on the double-slit demonstration. First, find the spacing between the slits. There are 527 lines/mm:

$$1 \text{ millimeter} = 10^{-3} \text{ meter}$$

therefore,

$$\frac{10^{-3} \text{ meter/mm}}{527 \text{ lines/mm}} = 1.89 \times 10^{-6} \text{ meter}$$

So the slit spacing is 1.89×10^{-6} meter. Putting this into the grating equation yields:

$$N\lambda = d \sin \theta$$

Transpose to:

$$\frac{\lambda}{d} = \sin \theta$$

$$\frac{(1)\ (633 \times 10^{-9})}{1.89 \times 10^{-6}} = 0.335$$

$$\text{Inverse sin } 0.335 = 19 \text{ degrees}$$

The first order line angular deviation is 19 degrees, compared to the 0.36 degrees previously obtained. Because the angle of deviation is much greater, the viewing screen can be moved much closer.

As an example, suppose you purchased a laser diode from a surplus house. The laser light from the diode is red, but since it was purchased surplus, you don't know the exact wavelength. To calculate the wavelength using the plastic diffraction grating, set up the laser under test in front of the grating and place your viewing screen 1 meter from the grating. Power up the laser, and measure the distance from the center fringe to the first fringe. Say you measured 0.394 meter. To calculate the frequency of light, start by using the tan function to calculate the angular deviation:

$$\frac{\tan 0.394}{1} = 0.394$$

$$\text{Inverse tan } 0.394 = 21.5 \text{ degrees}$$

$$N\lambda = d \sin \theta$$

Transposed is:

$$\begin{aligned} \lambda &= 1.89 \times 10^{-6} \times \sin 21.5 \\ &= 6.93 \times 10^{-7} \text{ meter} \\ &= 693 \text{ nanometer} \end{aligned}$$

Thin film interference

The vivid colors of the rainbow that are reflected from soap bubbles or from a thin film of oil on water are two examples of thin film interference. Here's how thin film interference fringes are made using a monochromatic light and two flat glass plates. Figure 11-22A shows the two flat glass plates illuminated with light from an HeNe Laser. The "air wedge" between the glass is exaggerated for illustration. In reality, the air wedge would be much smaller. The fringes created are the interference of the reflected light from the upper and lower sides of the air wedge. The bright fringes are naturally constructive interference. Here the two reflected waves are either in phase or an integral (whole number) number of wavelengths out of phase. The dark fringes represent destructive interference; the light here is ½ λ out of phase.

Notice that at the point of contact between the two glass plates, there is a dark fringe. It appears that at this point, where the air gap is extremely small and the two reflected light waves are traveling essentially the same distance, there should be a bright fringe. What is being over looked is when light is reflected from the surface of a material in which its speed is reduced, the reflected light is shifted half of a wavelength. That half-wavelength shift (λ/2, or 180 degrees) puts it out of phase with the other reflected wave; hence, a dark fringe.

When white light shines on a soap bubble, depending on the thickness of the soap film at any particular point, the reflected light interferes constructively or destructively. Figure 11-22B represents a section of a soap bubble.

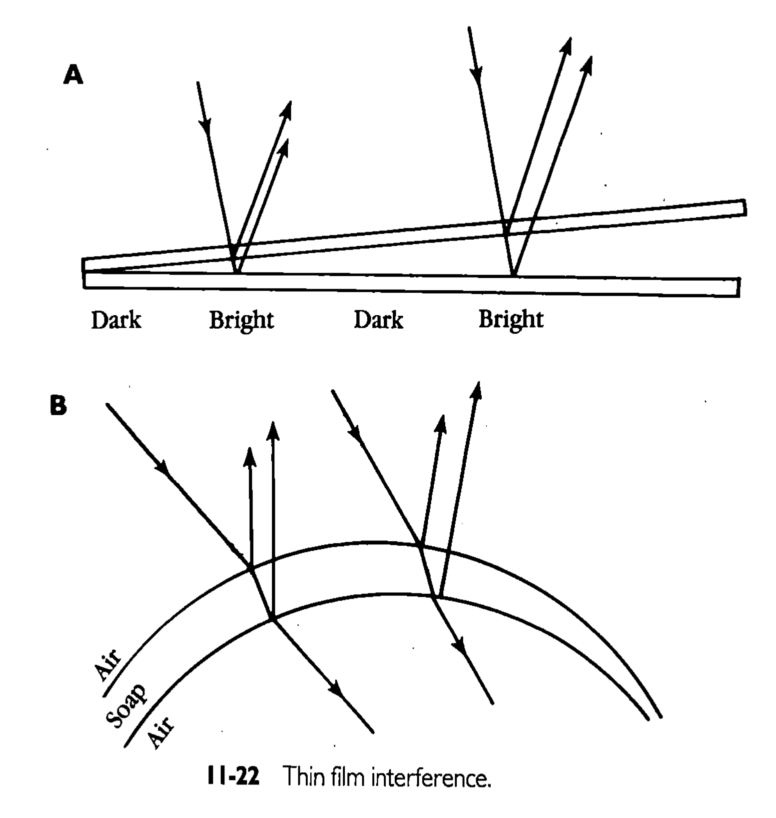

11-22 Thin film interference.

If the thickness of the soap film is such that wavelengths of reflected light are out of phase, destructive interference occurs. The colors representative of those wavelengths produce a dark fringe, and those particular colors are not visible. When the soap film thickness matches a wavelength of light so that the reflected light creates constructive interference, this gives rise to the vivid colors on the soap bubble film. Since the thickness of the soap bubble is usually varying, this produces the colors shifts and changes on the film surface.

NEWTON'S RINGS

The phenomenon of Newton's Rings is created when a plano-convex lens is placed in front of a flat glass and illuminated with monochromatic light (see FIG. 11-23). As shown in the previous example using the flat glass plates, light is reflected in basically the same manner.

Because the thickness of the air gap increases with distance from the center point of contact, the patterns of interference consists of light and dark concentric circles. Again, the air gap is exaggerated for illustration; in reality, it would be much smaller. If the setup is illuminated with light from an HeNe laser (633 nanometers), what is the thickness of the air gap at the seventh dark fringe? The center is dark due to the 180-degree ($\lambda/2$) phase shift of the reflected light (explained with the flat glass plates). Going out from the center to the first dark fringe, the light phase shift must be λ, which in turn means the

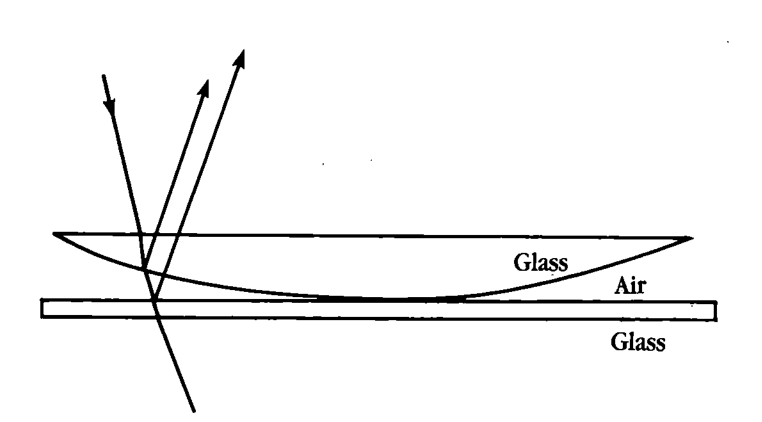

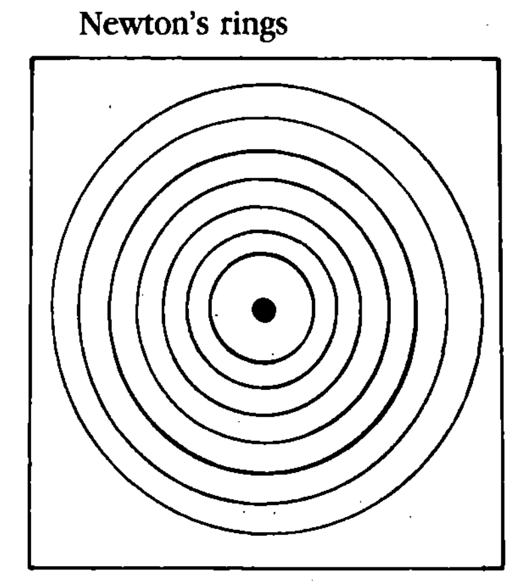

11-23 Newton's rings.

air gap there is half of that or $\lambda/2$ (remember that the light transverses the gap twice: first going down and then up). Continuing in this manner at the second ring, the air gap must have increased an additional $\lambda/2$. So at the seventh ring, the air gap would be:

$$\frac{7 \times (\text{wavelength})}{2} = \frac{7\,(633 \times 10^{-9})}{2}$$

$$= 2215.5 \times 10^{-9}$$

POLARIZATION

In the illustrations presented thus far, light has been depicted as a single waveform with the electric field going straight up and down and the magnetic field perpendicular to that. In reality, the electric field can be at any angle, but the magnetic field stays 90 degrees perpendicular to the electric field, regardless of the electric field's angle. To keep the following illustrations as simple as possible only the electric field of the light wave is shown.

Figure 11-24A shows an unpolarized beam of light. The electric and magnetic fields vibrate in all directions perpendicular to the direction of travel. Figure 11-24B AND C show polarized beam of light in which the vibrations occur in a single direction perpendicular to the direction the beams travels. This is called the *plane of polarization.*

Figure 11-25 illustrates how unpolarized light can be conveniently converted to polarized light by use of a polaroid sheet. A polarized sheet is a sheet of transparent plastic in which crystals of iodoquinine sulfate have been embedded and oriented. The resulting sheet allows light to pass through it only if the electric field is vibrating in a specific direction. If a second polarizing disk is placed in line with the first, the transverse wave nature of light can be demonstrated. If the axes of the two sheets are parallel, all the light passing through the first disk also passes through the second. If one of the sheets is rotated until its axis of polarization is perpendicular to the other, no light passes through the second sheet. This property is utilized in polaroid sun-

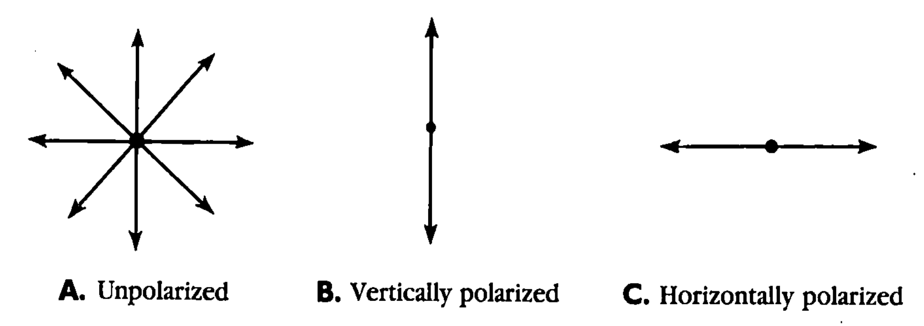

11-24 Polarized light. Imagine a beam of light is coming straight out of the page. The electric field vibration would then be as shown.

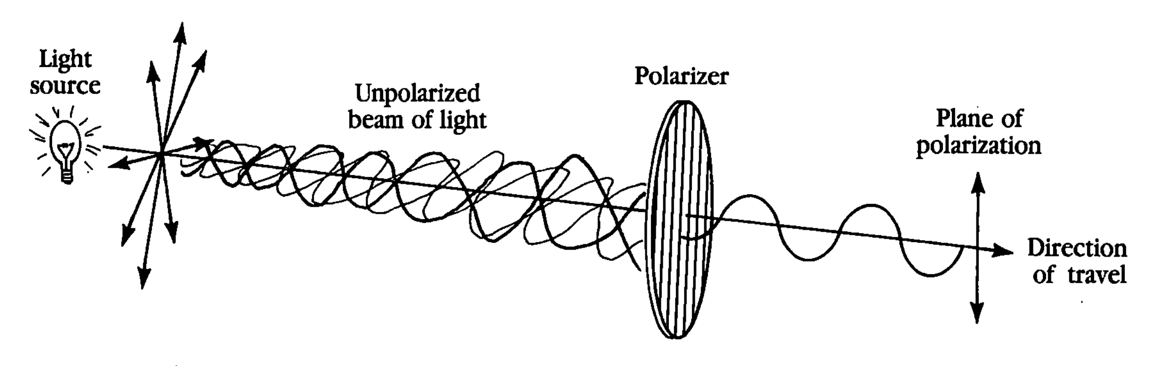

11-25 Use of a polarizing filter.

glasses. Figure 11-26 illustrates. Reflected light off a shiny surface is partially polarized in the horizontal plane. The polaroid material in sunglasses allows only vertically polarized light to pass through, thereby reducing glare. Since not all the light is polarized, you are still able to see.

For the holographer, polaroid material is very useful. With it you can check a laser to see if it's polarized, or you can use it to polarize a laser beam. In advanced holography setups when setting up a split-beam hologram, polaroid material is used to set a half-wave plate in the reference beam to ensure maximum interference and the brightest hologram possible. Fortunately, plastic polaroid sheets only cost a few dollars for a 9 × 12-inch sheet (see "Sources").

REFRACTION

Penetrating glass, air, or any transparent substance slows the speed of light appreciably for the time it travels through the substance. Once light leaves the substance, it resumes its original speed (see FIG. 11-27). If the light enters the medium at an angle, it is refracted. Refraction occurs because light travels at different speeds in the two media. The ratio of light speed in free space compared to light speed in a different media is called the *index of refraction,* which can be expressed as:

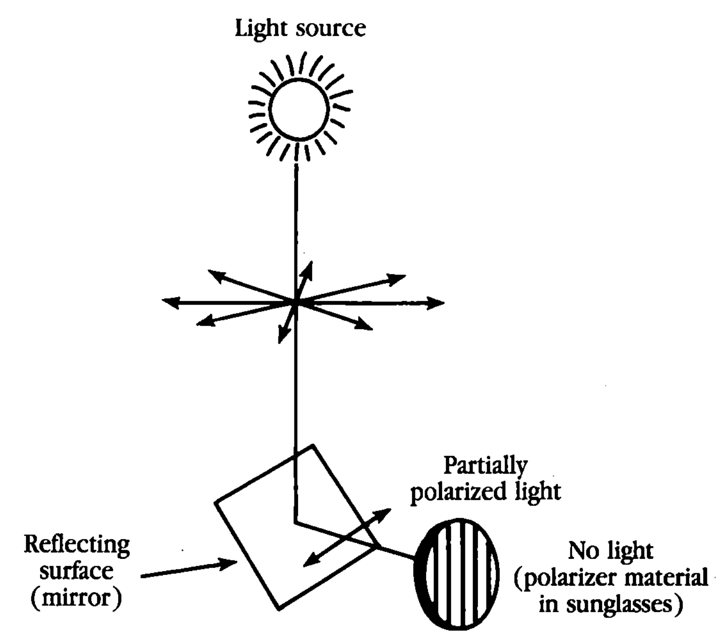

11-26 Operation of polariod-type sunglasses.

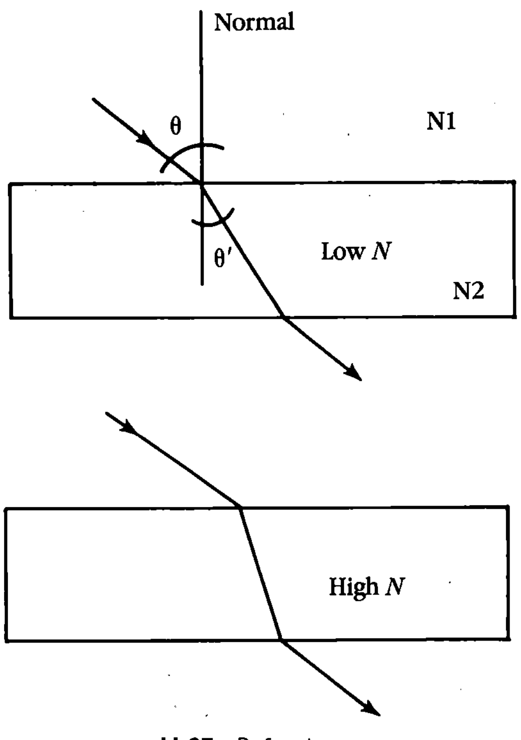

11-27 Refraction.

$$\text{Index of refraction} = \frac{\text{speed of light in vacuum}}{\text{speed of light in medium}}$$

$$n = c/v$$

where n = index of refraction
c = speed of light
v = speed of light in medium

Table 11-2 lists some materials and their indexes of refraction.

Table 11-2 Indexes of Refraction (n)

Substance	n
Air	1.0003
Water	1.3300
Lucite	1.5100
Glass, crown	1.5200
Glass, flint	1.6400
Diamond	2.4200

Snell's law of refraction calculates the angles of light entering a material.

$$n1 \sin \theta = n2 \sin \phi$$

As an example, suppose the angle of incidence θ is equal to 30 degrees. The light rays originate in the air ($n1$), which has an index of refraction of about 1. Assume it travels into glass with an index of refraction of 1.52. Then the angle ϕ can be calculated as:

$$n1 \sin \theta = n2 \sin \theta$$

$$\sin 30 = 1.52 \sin \theta$$

$$\sin \theta = \frac{\sin 30}{1.52}$$

$$\theta = \text{inverse sin } \frac{\sin 30}{1.52} = \frac{0.5}{1.52}$$

$$\theta = \text{inverse sin } 0.329$$

$$\theta = 19 \text{ degrees}$$

The ray enters the glass at 30 degrees from normal and exits the glass at 19 degrees.

VISIBLE SPECTRUM

The index of refraction varies slightly, depending on the frequency of light. Light with a higher frequency exhibits a greater value of n. For glass; the index of refraction for violet light is approximately 1 percent greater than red light.

This is succinctly demonstrated by a prism (see FIG. 11-28). A different value of n means a different degree of deflection when the light enters and leaves the glass. Therefore, directing a narrow beam of white light into one face of the prism separates into beams of various colors. This shows that white light is composed of many frequencies of light. The band of colors emerging from the prism is the visible spectrum.

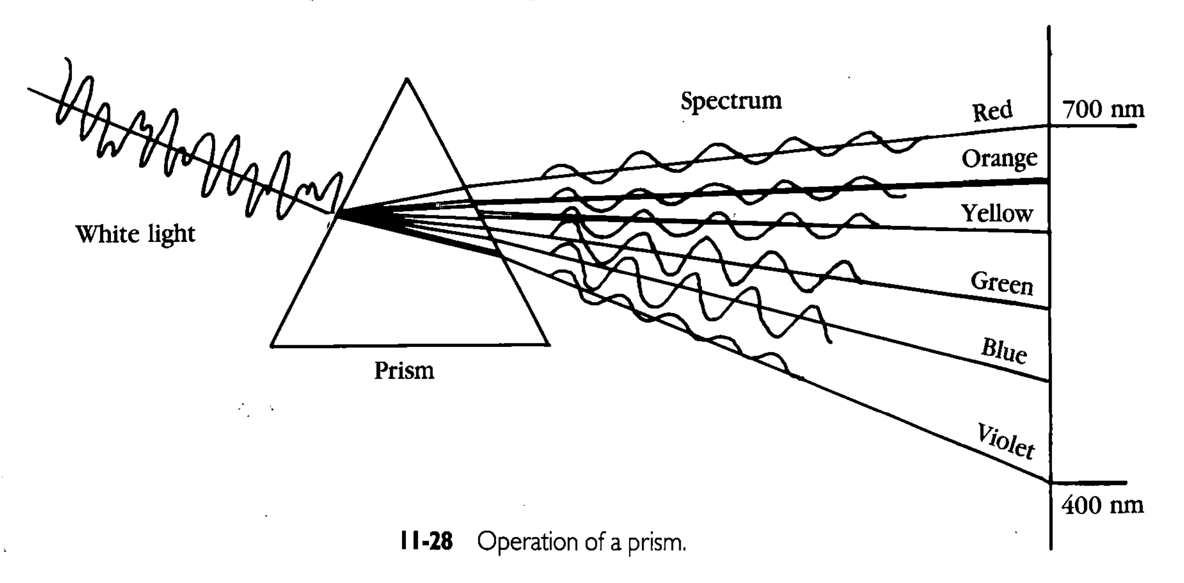

11-28 Operation of a prism.

FIBEROPTICS

A phenomenon of total internal reflection can occur when light passes from one medium to a second that has a lower index of refraction, for example from water to air.

Figure 11-29 traces four rays originating in water. Each ray has a different angle of incidence on the water-to-air interface. The angle of incidence for ray #2 is less than the angle of refraction, and the light ray is bent away from normal. As the angle of incidence increases, a critical angle is reached (ray #3) for which the angle of refraction is 90 degrees. This refracted ray travels along the interface of the two media. Increasing the angle of incidence further past the critical angle causes the ray to reflect back into the media as depicted by ray #4. To calculate the critical angle, use the refraction index of the two materials: air ($n1$) = 1.00 and water ($n2$) = 1.33. Thus:

$$\text{Critical angle (at ray \#3)} = \sin \frac{n1}{n2}$$

$$= \text{inverse sin } 0.752$$

$$= 49 \text{ degrees}$$

This is the basis of fiberoptics. The total internal reflection makes it possible to "pipe" light (see FIG. 11-30).

Figure 11-31 shows a cross-section for four types of optical fibers. N1 is the

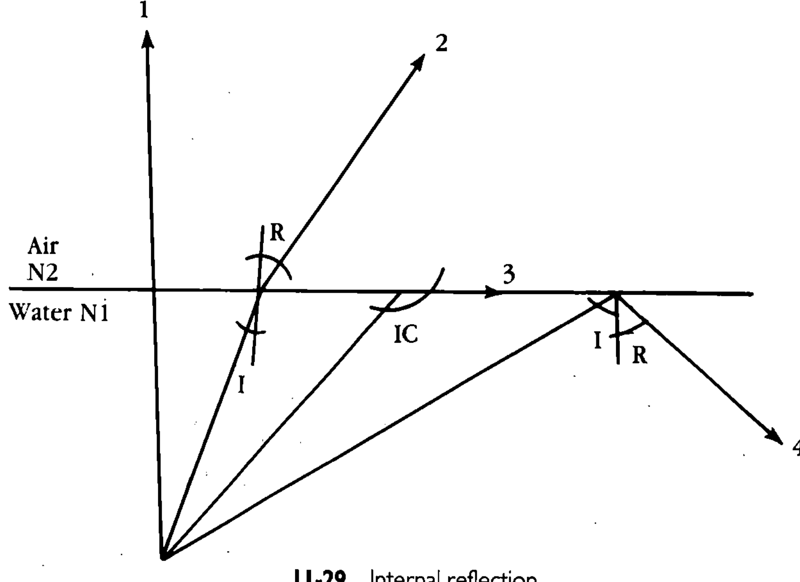

11-29 Internal reflection.

index of refraction for air, and N2 and N3 are the index of refraction for the fiber material.

A shows a common unclad fiber. B depicts a single-mode clad fiber. The index of refraction of $n2$ is greater than $n1$ but less than $n3$. Because of the small center core, this fiber is good for coherency light transmission. C is a multimode clad fiber suitable for incoherent light, and D is a single material fiber in which the outer shell isolates the core from outside disturbances. This minimizes cross talk in multifiber bundles. Cross talk is when a light signal leaks from one fiber to another and then transmits along that fiber.

Fiberoptics provide a new venue of techniques that promise to simplify difficult split-beam holographic setups. As FIG. 11-30 shows, light that enters a optical fiber transmits by bouncing internally, which would destroy a laser light's polarization and spatial coherency. The laser light emitted from this fiber is still useful for the illumination on the subject, but it cannot be used to transmit the reference beam.

More useful than this are the new types of fiber optics that maintain a

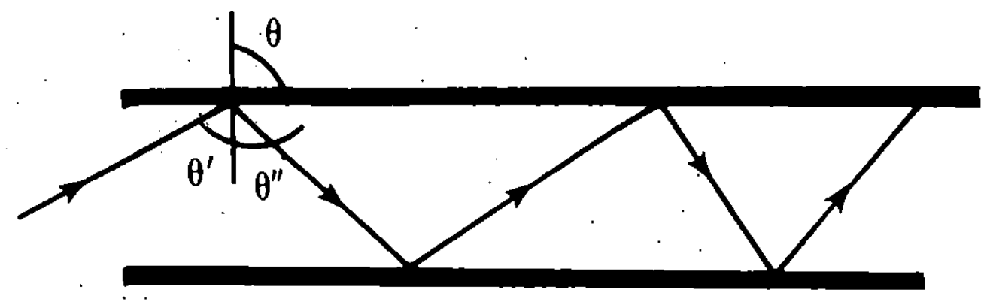

11-30 Internal reflection in a fiberoptic cable.

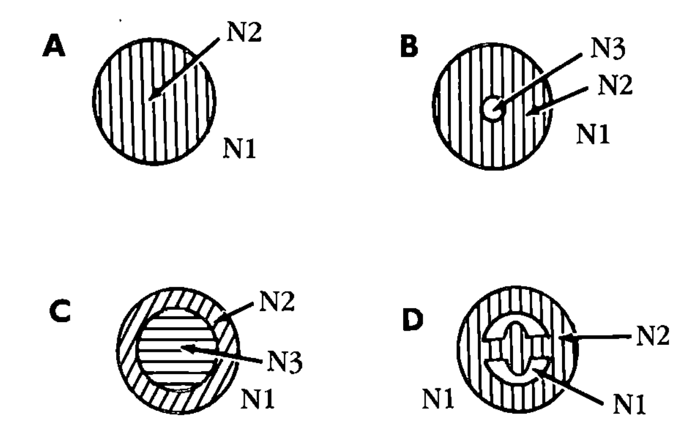

11-31 Cross section of various fiberoptic cables.

laser's coherency. These are monomode optical fibers similar to those depicted in FIG. 11-31B. To maintain coherency, the internal core must be extremely small, on the order of a few wavelengths across for the particular frequency in question. For HeNe lasers, the limit is 6 um or less. At this tiny diameter, the core acts as a waveguide to the laser light. Using this fiber permits "piping" both the object and reference beams. The most obvious use of fiber optics is to quickly match the optical path lengths to leave the holographer free to concentrate on the images. A second advantage is that the pathways are protected from air currents.

The main disadvantage of fiberoptics is the difficulty in directing the beam into the 6 um core of the fiber. The procedure is similar to directing a laser beam through a pinhole as when setting up a spatial filter (see Appendix B).

As a word of caution, when looking through various catalogs such as Edmund Scientific, you might run across advertisements for fiberoptics that specify coherency. In many cases, they are stating image coherence. In other words, an image can be transmitted through the fibers. Figure 11-32 shows a bundle of fibers in detail. The position of the individual fibers within the bundle is maintained throughout the fiber and subsequently can transmit an image. However, this is not the type of coherency required for working with holography.

THE ZEN OF SEEING

Light itself cannot be seen. I imagine many people just mentally protested that statement. The images we see with our eyes are the reflection of light off of objects into our eyes. Without an object to reflect light into our eyes, we could only observe the blackness of empty space, regardless of how much light might be present. This is why "outer space" is black. As an example, as you read these words on the page, you are looking at light reflected from the page to your eyes, but you can't see the light that is obviously present between the page and your eyes.

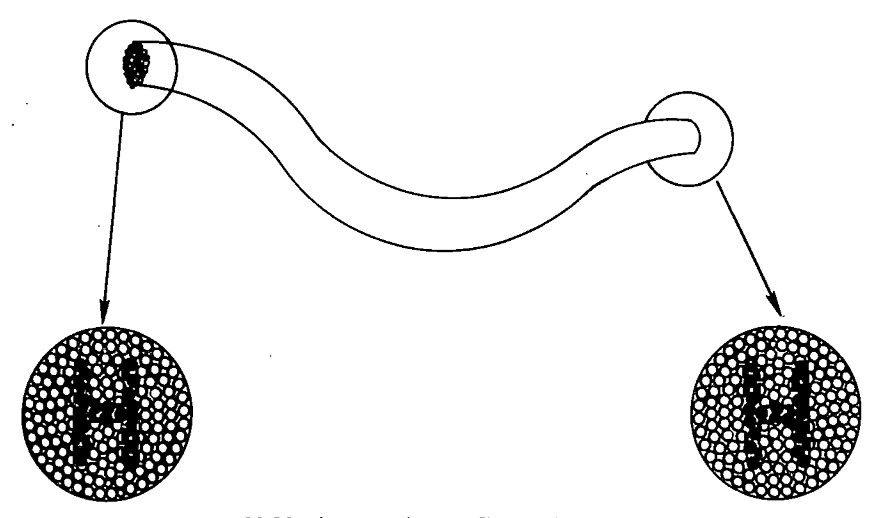

11-32 Image-coherent fiberoptic bundle.

Although it appears that we see objects in three dimensions, that is not really true. Each of your eyes picks up and relays to the brain a two-dimensional image. The brain performs a synthesis of the two images and gives the illusion of three dimensions. The interocular space between your eyes provides the necessary angular deviation of each image that makes this possible. This is also the underlying principle of the old stereophotographs and viewers that were the rage in the early 1900s. More recently, movies and pictures that use colored gel eyeglasses (3-D glasses) are a remake of this idea. Holographic stereograms also rely on the same principle.

Chapter 12

Laser Technology

Theodore Maiman, a scientist working for the Hughes Aircraft Company in California, won the race to make the first successful laser on May 16, 1960. He was considered a dark horse in the laser race because he continued to work with the synthetic ruby crystal as a lasing material, which was against the general belief that a gaseous material laser would be developed first.

The laser was a predicted derivative of the maser (*m*icrowave *a*mplification by *s*timulated *e*mission of *r*adiation), which had successfully been developed by Charles H. Townes, then at Columbia University. The word *laser* is an acronym for *l*ight *a*mplification by *s*timulated *e*mission of *r*adiation. Laser light is quite different from common everyday light.

ATOMS AND LIGHT

Although some of this material has been covered more completely in chapter 11, it is presented here again in an abbreviated form to paint a complete picture of laser technology.

All light is emitted from excited atoms. An excited atom is one that has absorbed additional energy. Typically, the energy absorbed by the atom could be from any number of sources such as heat, electrical discharge, collision, or absorption of a photon. The way atoms absorb and reemit energy can be described with a few basic rules. To examine how these rules operate, let's use the hydrogen atom as a model (the hydrogen atom is the simplest of all atoms, containing a single proton nucleus and a single orbiting electron). The light energy emitted from the hydrogen atom produces the simplest spectrum and is therefore the easiest to analyze. Figure 12-1 shows the possible energy levels of an electron in the hydrogen atom.

Whenever an atom absorbs energy, one of the atom's outer electrons jumps to a higher orbit. The energy levels are representative of the amount of

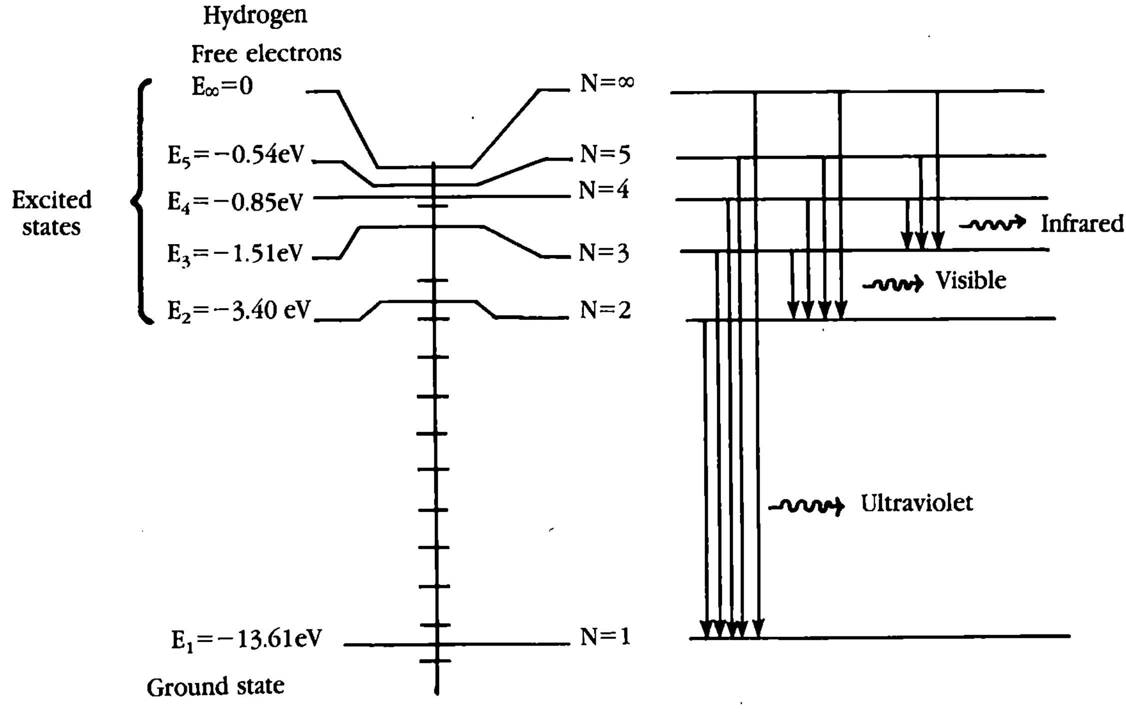

12-1 Energy states of the hydrogen atom.

energy the hydrogen atom can absorb and the electron jump when a hydrogen atom absorbs the energy. Each level's energy is also indicative of the energy emitted when an electron falls from a higher level to a lower level.

The energy levels of the electrons are shown as negative numbers. The negative sign illustrates that the energy must be added to the atom. It might not be obvious, but the energies represent the energy required to remove the electron from the hydrogen atom at various shell levels. This is called *ionization energy,* and for hydrogen in the ground state, it is 13.61 eV. The lowest level in the diagram represents this normal state or ground state of the atom ($n=1$). The next level up is the 10.21 eV above-ground state and is the first excited state. This also illustrates the minimum amount of energy required to raise hydrogen to the first excited state. As an example, if an electron with 8 eV of energy collided with the hydrogen atom, it would scatter away from the atom and not be absorbed. If on the other hand the electron contained 10.21 eV or greater energy and collided with the hydrogen atom, the energy could be absorbed. Any excess energy stays with the electron and is carried away as it leaves. The now excited hydrogen atom would have its electron in the $n=2$ shell. Typically, an atom in an excited state can only remain excited a short amount of time, 10^{-8} second before reemitting the excess energy. The excess energy in the example atom above for instance would reemit a 10.21 eV packet of electromagnetic energy.

The amount of energy the atom releases in eV can be used to determine the wavelength of the emitted radiation. Rather than go through a series of complex equations to find an answer, use instead the following basic rule of thumb:

$$1 \text{ eV} = 1242 \text{ nm}$$

To calculate the wavelength of an emitted photon, the first step is to find the energy released in the electron jump (transition). For the hydrogen atom:

$$n_2 - n_1 = \text{eV}$$

$$-3.40 \text{ eV} - (-13.61 \text{ eV}) = 10.21 \text{ eV}$$

So the hydrogen atom releases a quanta of energy equal to 10.21 eV. To convert this answer into the wavelength of the emitted energy:

$$1 \text{ eV} = 1242 \text{ nm}$$

$$\frac{1242 \text{ nm}}{10.21 \text{ eV}} = 121.64 \text{ nm}$$

This answer corresponds almost exactly to the actual wavelength emitted. You can use this rule of thumb to calculate wavelength emitted or absorbed for any electron transitions.

White light

White light is composed of all the colors of the rainbow. These visible wavelengths range from 400 nm to 700 nm. It would appear from the way light is generated from atoms that only specific colors or wavelengths should be emitted, depending upon the type of atom emitting the light.

Returning to Thomas Edison's incandescent lamp from the last chapter, this type of lamp produces a continuous visible spectrum from 400 to 700 nm. Why doesn't this lamp emit the single wavelengths associated with the element tungsten, from which the filament is made? The short answer is that when atoms are in close proximity to one another, such as in a solid, liquid, or compressed gas, the electrons in the outer shell of the atoms mesh together, allowing electrons to jump from one atom to another. As a result, a photon of essentially any energy can be emitted in one electron jump or another.

Stimulated emission

Atoms can gain energy to become excited from a number of different sources: heat, electric current or discharge, collisions, etc. Another way, and one that is of special importance to us, is through the absorption of a photon of light. If the energy of the photon matches the difference in energy between two electron levels of an atom (the level it's in and the level to which it can jump), it will be absorbed (also see chapter 11).

In 1917, Albert Einstein published a paper on stimulated emission of light. This paper proved to be the foundation upon which laser technology is built. As we have seen, an atom in the excited state reemits its excess energy in approximately 10^{-8} seconds. This process is called *spontaneous emission, spontaneous* meaning it could occur at any time, but usually within the 10^{-8}-second period. Einstein showed that the emission of the photon could be stimulated to occur by another photon. This happens when an excited atom

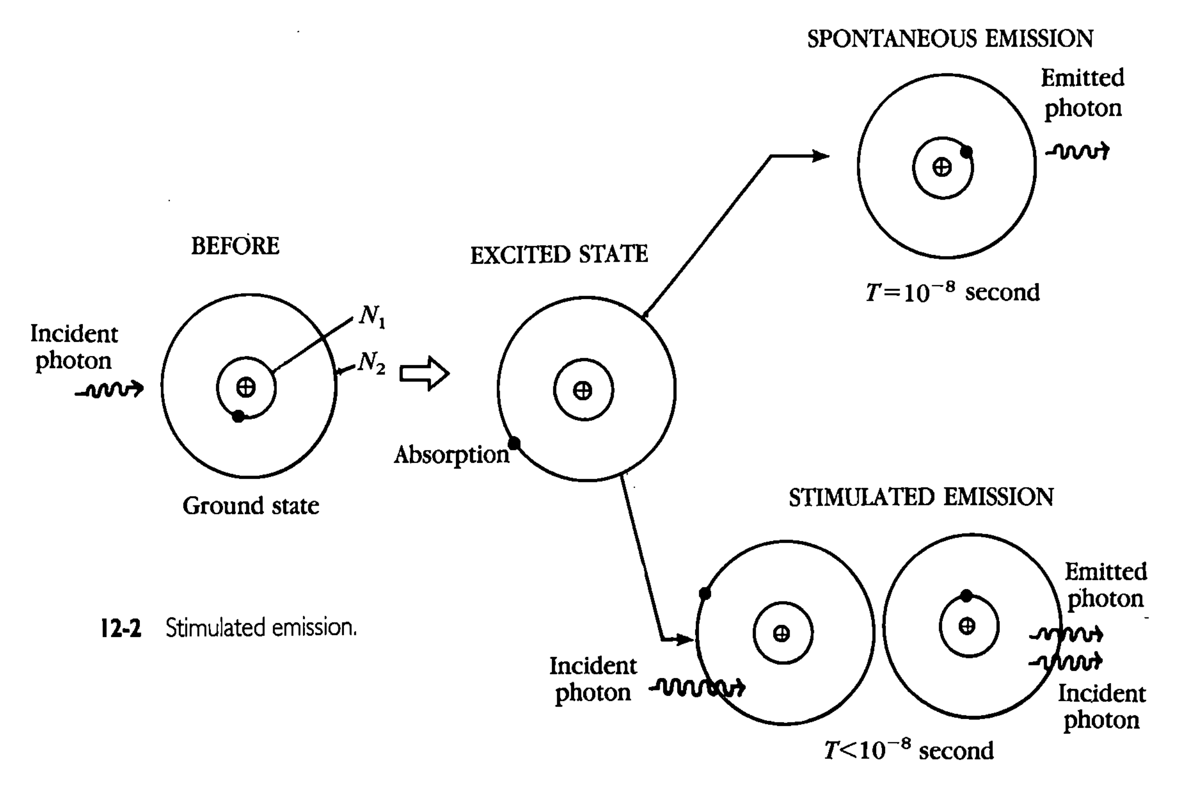

12-2 Stimulated emission.

collides with an outside photon with the same energy as the one to be emitted from the atom. Amazingly, the two photons (the original one and the stimulated one) leave the atom together, traveling in the same direction and exactly in phase with one another. This is called *stimulated emission* and is illustrated in FIG. 12-2. Stimulated emission was clearly seen as a way to amplify light, but it wasn't until almost 40 years later that this principle could be put into practice.

Metastable states One of the drawbacks with stimulated emission is the quickness with which excited atoms decay and undergo spontaneous emission. This doesn't leave sufficient time for much stimulated emission to occur, so most of the energy absorbed in this type of material would still be radiated spontaneously. Certain excited states, however, are *metastable* (temporarily stable) and can last much longer (100,000 times longer), approximately 10^{-3} seconds. The operation of the laser depends upon these metastable states in atoms because the metastable states allow sufficient time for stimulated emission to amplify light. Figure 12-3 shows an electron of an excited atom spontaneously decaying from an excited state, to a metastable state, then to ground state.

Population inversion There is a subtle point to be made with stimulated emission that isn't always obvious. Looking back at FIG. 12-2, imagine that either of the two photons leaving the stimulated atom could be easily absorbed if it collided with another atom in the ground state. This absorption would quickly quench any light amplification. In other words, absorption and stimulated emission are competing processes. In order for stimulated emission to amplify light, there must be more excited atoms to contribute photons than unexcited atoms to absorb them. The process to create more excited atoms than unexcited atoms is called *population inversion.*

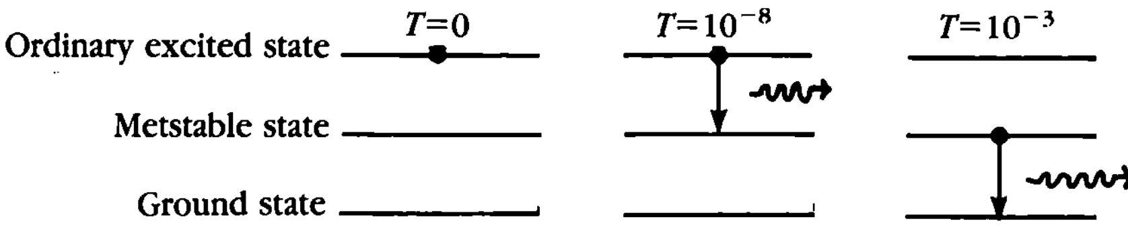

12-3 Metastable states.

RUBY LASER OPERATION

With the basic processes explained, let's tie them together to produce a working model of the first successful laser made, the ruby laser. The ruby rod material is an impure aluminum oxide (Al_2O_3) where a few of the aluminum ions (Al^{+++}) in the crystal are replaced with chromium ions (Cr^{+++}). The chromium is responsible for the ruby's reddish color. The percentage of chromium in the crystal material is low, typically 0.5 to 1 percent. Chromium has the ability to absorb light and scuttle into a metastable state before reemitting its absorbed energy. If a low level of light is used, an insufficient amount of chromium ions would be excited. To create a population inversion, a flash lamp is used to produce a brilliant pulse of light that raises most of the chromium ions to the excited state (see FIGS. 12-4 and 12-5). The excited chromium ions quickly decay to their metastable state. Because the metastable state lasts so long (in atomic time), the chromium ions accumulate here and produce a population inversion. As the first chromium ions in the metastable E2 level drop to ground state, they emit a photon of light. As these photons hit other excited chromium ions, stimulated emission occurs. Mirrors are at the ends of the ruby rod which reflect the light traveling parallel to the axis of the rod. As this reflected light transverses the ruby crystal again and again, the light amplifies and builds in intensity through stimulated emission of excited atoms on each pass, quickly building to an intense pulse of laser light that emerges from the partially silvered mirror—all in approximately 10^{-6} seconds.

The distance between the mirrors is critical. It is made to be exactly a whole number of wavelengths apart so it forms an optical resonance cavity where the light waves bouncing between the mirrors form an optical standing wave. It is this standing wave pattern that is responsible for keeping the stimulated photons in step, which is the reason the pulse of laser light that emerges from the rod is both coherent and collimated.

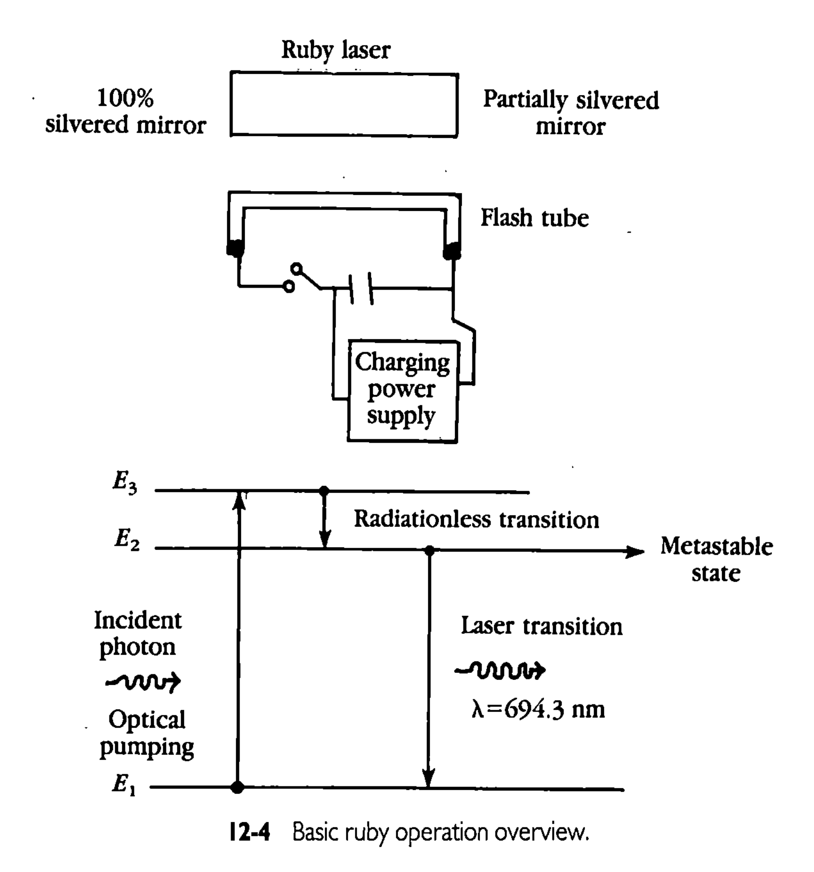

12-4 Basic ruby operation overview.

Light losses in the ruby laser

The ruby laser operation just described is an idealized version of a laser. The laser isn't perfect; there are some light losses that should be mentioned. First, any light emitted from the chromium ions that aren't in proper alignment with the mirrors escape through the walls of the rod. Any stimulated emission caused by these escaping photons is also lost.

The second loss is at the partially silvered mirror. On each pass, a small percentage of light escapes from the partially silvered mirror. Some people wrongly believe that the mirrors reflect all of the light until the intensity builds to a point where the partially silvered mirror magically transmits the laser beam. This is not so; some light is lost on each pass. The laser beam output power is only a small percentage of the total beam power existing within the ruby rod. The output power is proportional to the percentage of light that escapes from the partially silvered mirror.

One final consideration are the chromium ions that reemit to the ground state without stopping at the metastable state. This light doesn't contribute to the laser function, but the chromium atom in the ground state can be reexcited.

Multipulse

Once the ruby rod is operating as a laser, the E_2 metastable state quickly depopulates by stimulated emission in approximately 10^{-6} seconds. Because

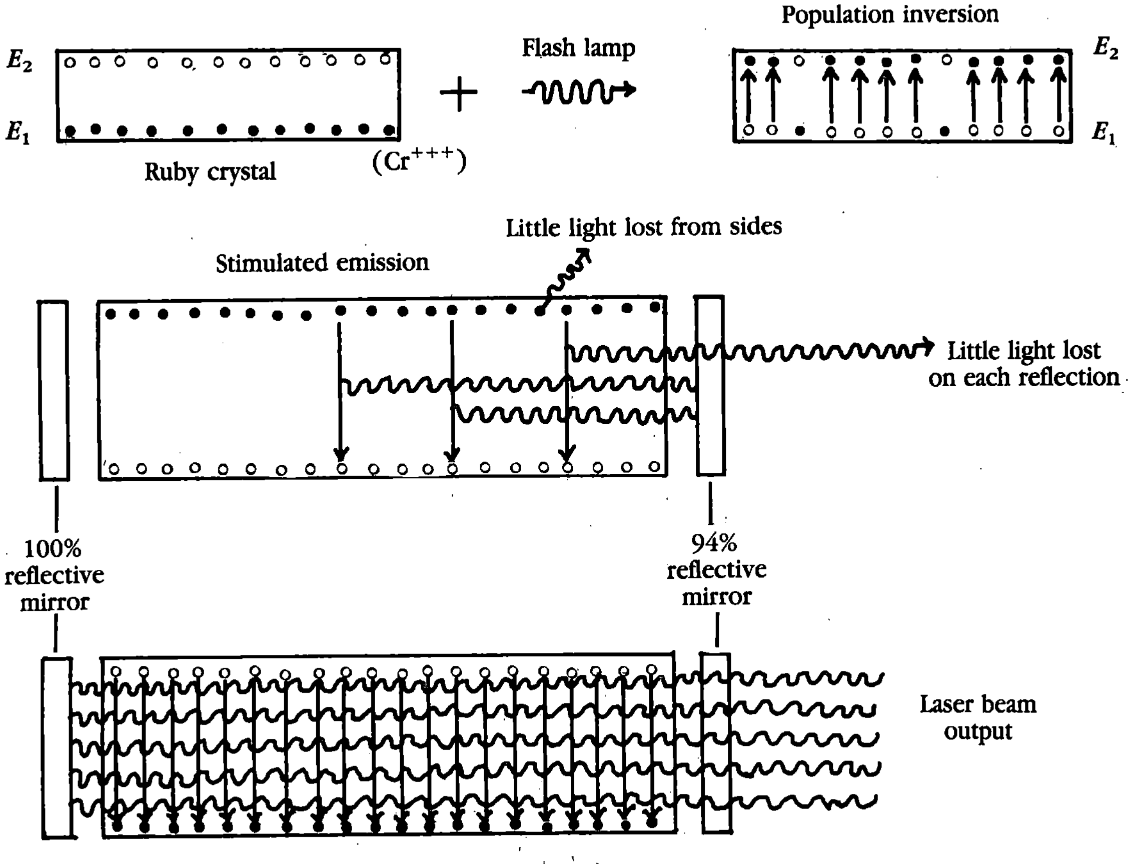

12-5 Ruby rod operation.

the output pulse from the flash tube lasts much longer than the laser pulse, more than one population inversion occurs in the ruby rod. This produces a series of irregularly spaced laser pulse spikes about 10^{-6} second in duration in an envelope defined by the flash lamp discharge (see FIG. 12-6).

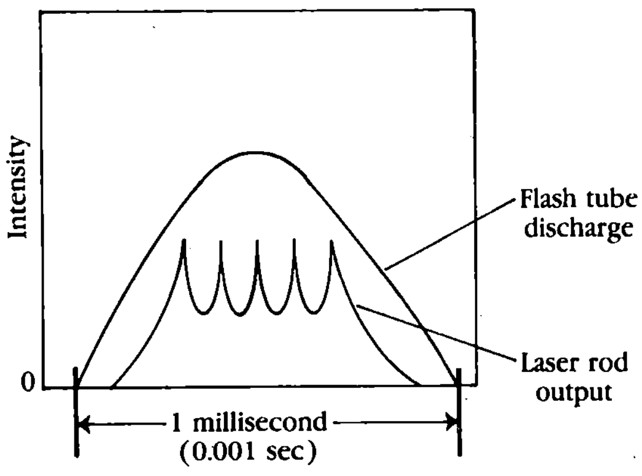

12-6 Multiple-pulse characteristic of ruby laser.

Q-switching

The output pulse of the ruby laser can be controlled by a process known as Q-switching. Q-switching can confine the output of the ruby laser to a single pulse with a very short period (25 to 50 ns) and higher energy. This can be achieved with either of two Q-switching processes called *passive* and *active* switching. Both systems involve blocking one of the cavity mirrors. When a cavity mirror is blocked, light cannot oscillate between the mirrors, and laser action is effectively quenched. When the flashtube is fired with the cavity mirror blocked, the population inversion proceeds further than it would under normal operation. Because the reflections are blocked, the light isn't bouncing back and forth, depopulating the metastable state through stimulated emission. So a greater number of atoms accumulate in the metastable state, and this essentially stores greater energy in the ruby rod. By quickly unblocking the mirror, the optical cavity is restored, and laser operation begins. The peak output power of the laser is much greater because of the stored energy in the metastable state.

In a passive system, a dye such as rhodamine is used. A property of the dye molecules is that they are capable of absorbing a photon of laser light, after which they become transparent to any additional laser light and are said to be *bleached*. To use the dye as a Q-switch, place the dye between one of the cavity mirrors and the laser rod. At first, the dye molecules are opaque to laser light. When the flash lamp fires, the dye absorbs photons of light from the stimulated emission in the laser media. The dye bleaches almost instantaneously, quickly changing from an absorption to transmissive state, exposing the mirror, and restoring the optical cavity. The laser pulse is generated and emitted. The dye quickly releases its energy and returns to it absorptive, opaque state.

An active Q switch is like an electronic shutter, similar in operation to an LCD panel (see chapter 11). These Q switches are representative of a number of different techniques that accomplish the same thing such as cavity dumping and modelocking.

HELIUM-NEON LASERS

The first successful gas laser was made a few short months after Maiman's ruby laser. Ali Javan of the Bell Telephone Labs and two colleagues, William R. Bennett Jr. and Donald R. Herriott, demonstrated the first laser in late 1960. The first successful HeNe laser built by Javan wasn't visible to the human eye. It operated in invisible infrared region of the spectrum. Since then, many types of gases have been used successfully as host materials in lasers; however, this book concentrates on the helium-neon laser because it is the most commonly used laser in holography, and its principles can be applied to many other gas lasers.

The helium-neon (HeNe) laser operates on the same basic principle as the ruby laser. It uses a gaseous host material instead of a ruby crystal. The gas is a mixture of nine parts helium to one part neon at low pressure in a glass tube. The glass tube has mirrors at both ends, forming an optical cavity. An electric discharge ionizes the gas. The helium atoms absorb energy from the discharge and jump to one of two possible metastable states identified as $2S_1$ and $2S_2$ in

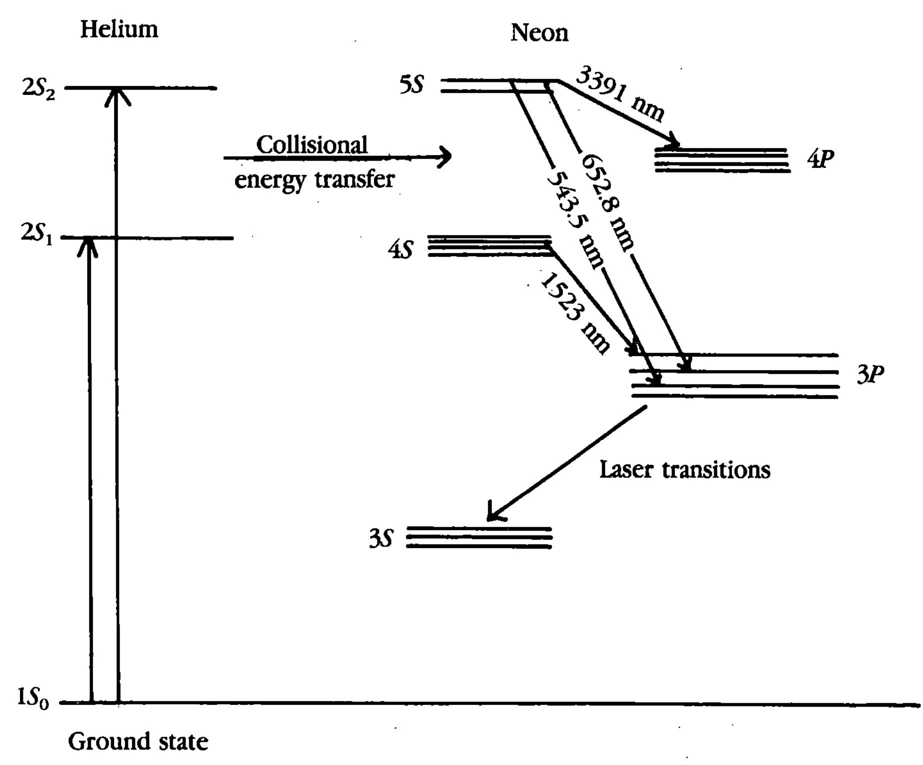

12-7 Excited energy levels of hydrogen and neon.

FIG. 12-7. The energy level of excited helium matches the metastable states of neon. So whenever an excited helium atom collides with a neon atom in ground state, it can transfer all of its energy to the neon atom. Because there is so much more helium than neon in the glass tube, the helium essentially keeps the neon excited and in a population inversion state.

When the neon absorbs this energy from the helium, one of the neon atom's outer electrons jumps to either the 4*S* or 5*S* energy level. Figure 12-7 shows that there are numerous downward energy jumps from these levels for an excited neon atom to make. The energy jumps are more complex than the hydrogen atom because neon has a more complex atomic structure. Neon can be made to operate as a laser at a number of different wavelengths, but the most familiar wavelength is 632.8 nm, the standard red HeNe laser. But an HeNe laser can operate at 543 nm, emitting a green laser light. Green HeNe lasers and/or tubes are available from Melles-Griot company.

Optical cavity

The mirrors at the ends of the HeNe tube form an optical cavity that reflects the light back and forth, amplifying the light with stimulated emission on each pass, as in the ruby laser. A major difference between the ruby and gas laser is that the HeNe laser operates continuously, providing a continuous wave (cw) laser beam. Again, the two mirrors are a whole number of wavelengths apart to form a resonant optical cavity that produces a standing optical wave pattern. Only the wavelengths than can form a standing wave pattern between the mirrors will be amplified. Other wavelengths will be reflected out of phase, undergo destructive interference, and cancel out. One of the cavity mirrors allows a small portion of the laser light to be transmitted, usually about 1 percent. It follows then that the internal beam of the laser is about 100 times more powerful than the output beam.

The gas laser doesn't use a flat mirror optical cavity as with the ruby rod (see FIG. 12-8A). The flat mirror configuration is harder to align and is somewhat unstable in the gas laser. Other optical cavity configurations are employed that allow easier mirror alignment and enhanced stability. Figure 12-8B shows a confocal optical cavity that is constructed of two concave mirrors of equal radius. This system is less sensitive to alignment but has a drawback of being quite sensitive to mirror separation. Figure 12-8C is the more practical system. From the drawing you can see that only about one third of the plasma is being utilized to generate the laser beam. This power limitation drawback is more than compensated for by the ease of alignment and stability.

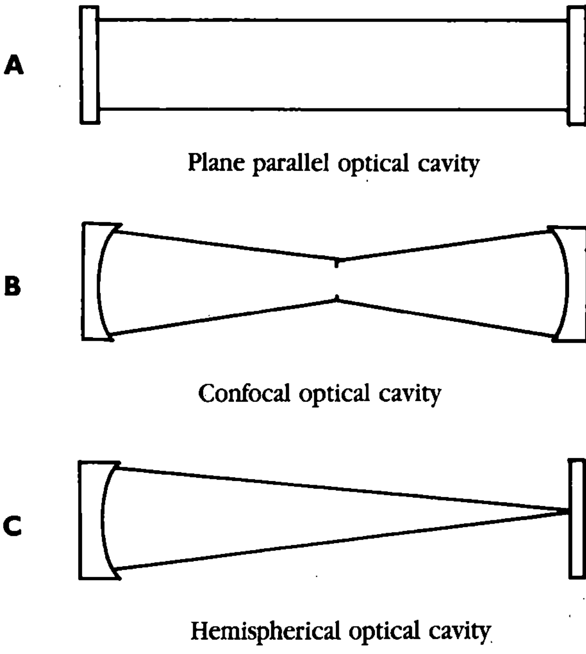

12-8 Optical cavity configuration for gas lasers.

Longitudinal modes

As stated earlier, the mirror separation must be a whole number of wavelengths apart. This produces a standing wave pattern of light. Any light that isn't an exact frequency multiple of the length will be reflected out of phase and cancel out. The distance our light travels is $2L$, where L = length of the mirror separation, so:

$$2L = m\lambda$$

where m is a large whole number.

If the cavity length is 25 cm, then the round-trip distance the laser beam would travel is $2L$ or 50 cm. The length of 50 cm is equal to about 790,000 wavelengths of an HeNe laser's 632.8-nanometer light. However, resonance not only occurs at 790,000 waves, but also at 790,002, 790,004, etc. These other wavelengths can also oscillate within the laser cavity. These additional wavelengths are very close to 632.8 nanometers and are called the *longitudinal modes* of the laser. All wavelengths that can oscillate in the optical cavity, along with the Doppler broadening (explained later in this chapter), taken in total, define the bandwidth of the laser light emitted from the laser tube. Figure 12-9 shows the primary resonance line of 632.8 nm along with four longitudinal modes or additional cavity resonances. The bandwidth of the laser is calculated at the *f*ull-*w*idth *h*alf *m*aximum (FWHM) point.

Be aware that for these longitudinal modes to be amplified, their individual frequencies lie within the finite spectral line width of the atomic transitions of neon. This information clearly shows that lasers are not perfectly monochromatic or perfectly coherent as commonly thought. Even so, today's off-the-shelf lasers provide a source of coherent light that is hundreds of times longer than was formerly possible.

Coherence length and temporal coherence

Lasers have two types of coherency: *temporal* and *spatial.* The temporal coherence refers to the length two light waves remain in phase as they travel outward. The additional longitudinal wavelengths emitted by the laser limit the coherency length of the laser. This is illustrated in FIG. 12-10, which shows two wavelengths emitted from the barrel of the laser tube. One wave is the primary line of the HeNe laser, 632.8 nm. The other represents a longitudinal line at 632.8006 nm. As the waves leave the barrel at point A, they are in phase. As they travel away from the laser, the tiny difference in their wavelengths causes the wavelengths to shift out of phase with one another, as shown at point B. Continuing further, the waves realign in phase at point C. Continuing further

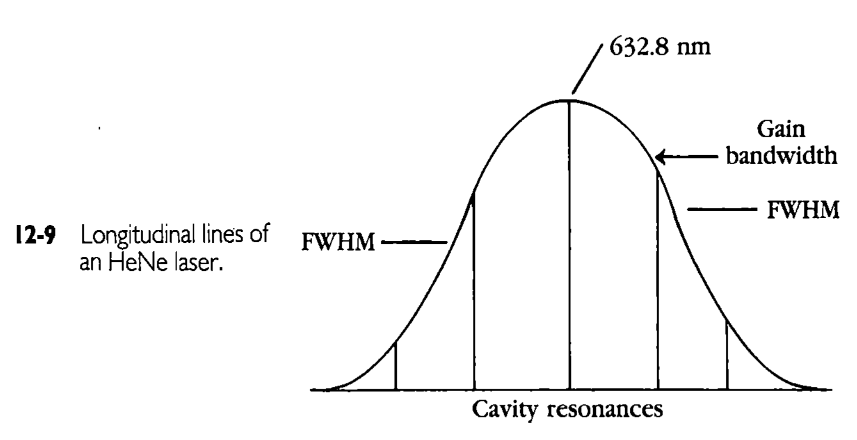

12-9 Longitudinal lines of an HeNe laser.

from this point, the process repeats itself. If the two wavelengths were summed to show interference, there would be constructive interference at point A, gradually reducing to destructive interference at point B, and rebuilding back to constructive interference at point C.

The coherence length of the laser can be defined as the minimum distance traveled by two wavelengths where optical interference vanishes and cannot form interference fringes. Looking back at FIG. 12-10, it might appear that the coherence length of the laser should be defined at the distance between points A and B. It appears this way because the wavelength is not shown to the left of point A. To clarify the point, cut and paste the waveform from point B to C and place it on the left side of point A so that points C and A are aligned together. You will see that the coherence length hasn't changed.

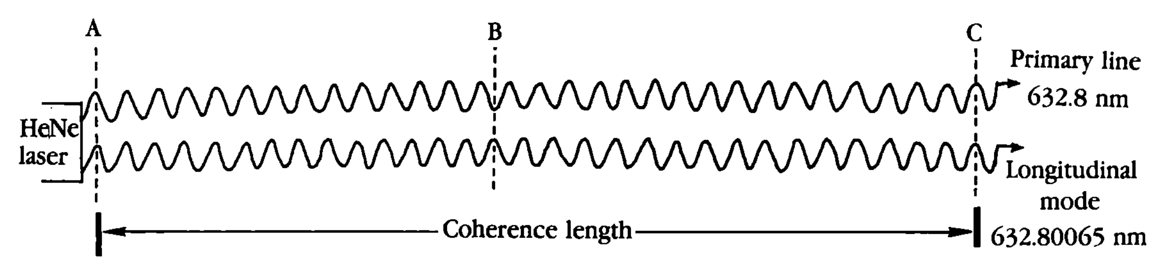

12-10 Coherence length of HeNe laser. The laser light waves emitted following C are the same wavelength and in phase with one another.

The coherency length of the laser determines the maximum depth with which you can produce a hologram. In a split-beam setup, if the difference in the beams' optical paths equals or exceeds the coherence length, the primary line of the laser and the longitudinal lines are out of phase and the beams do not form interference fringes, preventing a hologram from being recorded on the film emulsion. This is why the laser beam paths are equalized in split-beam setups to try to ensure the production of the interference pattern by the two beams. (The beam path measurements are made from the beam splitter.)

This facet of optics is used in a Michelson interferometer. With the Michelson interferometer, the coherence length of the laser can be physically measured. In addition, the stability and relaxation time of the isolation table can also be checked (see chapter 2).

Calculating the coherence length Aside from physically checking the coherence length of the laser using the Michelson interferometer (which is the best way), the coherence length can be calculated using either the range of frequencies (ΔV) or the range of wavelengths ($\Delta\lambda$). With the range of frequencies, use the equation:

$$\text{Coherence length} = c/(2\Delta V)$$

If you know the wavelength range, use the equation:

$$\text{Coherence length} = \lambda^2/\,(2\Delta\lambda)$$

EXAMPLE:

Suppose you purchase a visible laser diode that emits light at 670 nm and has a wavelength range of 0.2 nm. To calculate its coherence length, plug this information into the equation.

$$\text{Coherence length} = (670 \times 10^{-9})^2 / 2\,(.2 \times 10^{-9})$$
$$= 0.0011 \text{ meter}$$

which can be converted to 1.1 millimeter or approximately 1⁄20 inch. This laser would appear to only be useful for creating very flat holograms.

Many HeNe laser manufacturers give the bandwidth of the longitudinal modes of their tubes. A typical bandwidth at FWHM is 1500 MHz (megahertz). Plugging this information into the appropriate equation:

$$\text{Coherence length} = 3 \times 10^{8} / (1500 \times 10^{6})$$
$$= 0.2 \text{ meter}$$

which can be converted to 20 cm, or approximately 8 inches.

Doppler broadening

Doppler shift is the effect when you hear the change in pitch of a police car or ambulance siren as it rushes past. When the vehicle is moving toward you, your ears receive more sound waves per second than the siren is actually putting out. This is perceived as a higher pitched sound. When the vehicle passes and is moving away, you hear a lower pitch. This same effect occurs with electromagnetic waves. The famous astronomer Edward Hubble used the Doppler shift of light to determine the relative speed of stars in relationship to the earth. Stars moving toward us had their spectral lines of hydrogen shifted to the blue end of the visible light spectrum and became known as the *blue shift*. Stars receding from earth had spectral lines shifted toward the red end of the visible spectrum and became known as the *red shift*.

The gas atoms in a laser tube are in constant motion. This motion creates a Doppler shifting of the laser light, which essentially broadens the spectral line of the atomic transitions. Looking back at FIG. 12-9, the gain bandwidth curve contains and defines the Doppler-shifted laser light emitted from an HeNe laser.

TEM modes and spatial coherence

TEM stands for *t*ransverse, *e*lectric and *m*agnetic modes. We have already seen how the length of a laser's optical cavity establishes longitudinal modes. TEM modes represent the electric and magnetic field vectors across the diameter of the laser tube. These transverse modes are designated with two subscript numbers m and n. The subscripts refer to the number of nulls (zero-intensity points) across the diameter of the tube for the electric and magnetic fields. The electric and magnetic fields are perpendicular to one another. Figure 12-11 shows the laser beam's energy distribution for various TEM modes. The figure

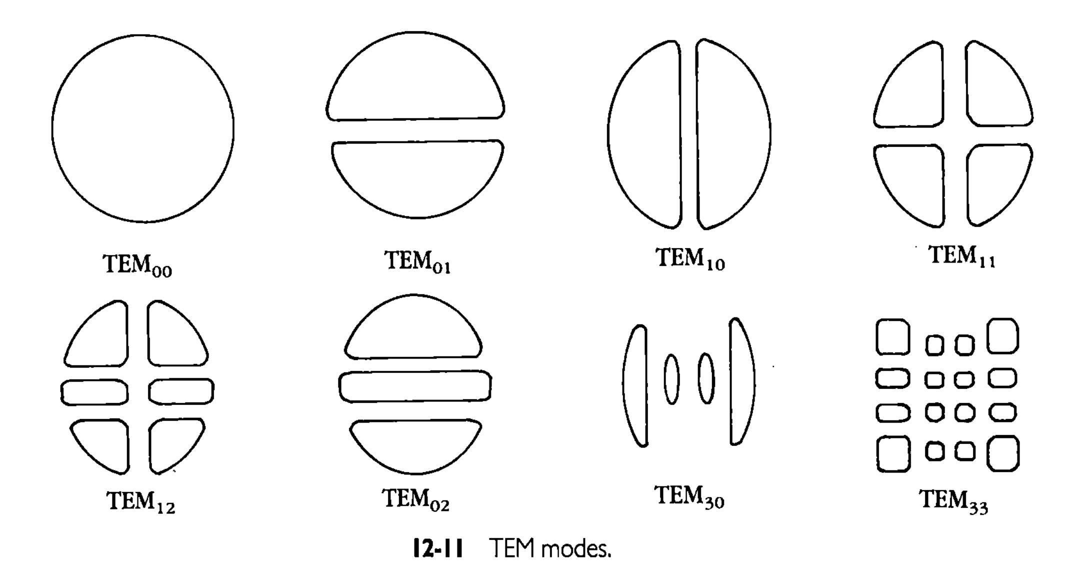

12-11 TEM modes.

shows that at TEM_{01} and TEM_{10}, the beam has a single wall that divides the beam into two. TEM_{11} has two nulls in each direction, which divides the beam into four.

Holography requires a TEM_{00}-mode laser. This mode provides the smoothest energy distribution of the laser across the beam's diameter, minimum diffraction losses, minimum beam divergence, and it can be focused to the smallest possible spot. The TEM mode also defines the spatial coherence of the laser. TEM_{00} is spatially coherent. The spatial and temporal coherence of a laser are independent of one another.

Fabray-Perot etalon

We have looked at different characteristics of the HeNe laser that broaden the wavelengths emitted by the laser. There are ways to limit the bandwidth of the laser to provide a narrower line. The most common technique employed to achieve this is a Fabry-Perot etalon. The *etalon* is a pair of optically flat reflective surfaces inserted between the laser mirrors. See FIG. 12-12A and B. The etalon itself is a secondary resonant optical cavity. At most wavelengths, the etalon either reflects the light away from the main cavity or is canceled through destructive interference to suppress any laser oscillation at those wavelengths. At certain wavelengths, the reflections are nullified, allowing the light to pass through unobstructed and laser operation is sustained. The etalon can be tuned to different wavelengths by either adjusting the spacing of the optically flat surfaces or by tilting the device in the optical cavity.

By removing the longitudinal modes, the etalon effectively increases the coherency length of the laser, making it possible to produce holograms of greater depth. As an example, suppose the etalon reduced the laser bandwidth to 1 megahertz. Plugging this value into the equation we have:

$$\text{Coherence length} = 3 \times 108 / 1 \times 106$$
$$= 300 \text{ meters}$$

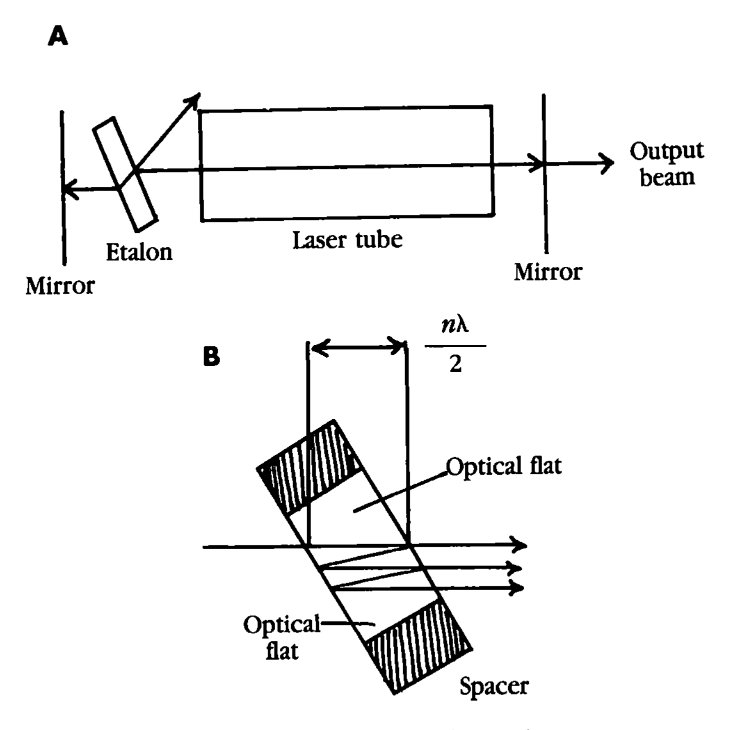

12-12 Etalon placement and operation.

HeNe laser beam divergence

The beam divergence from the HeNe laser depends in part upon the type of optical cavity used. When you purchase a laser or a laser tube, the beam divergence of the laser is given. Typically, this value is about 1 milliradian. A radian is equal to 57.29 degrees; therefore, a milliradian is equal to 0.05729 degrees. Use simple trigonometry, to calculate the beam spread (FIG. 12-13 illustrates). The equation is:

$$D = 2L \cdot \sin \theta$$

where D = beam diameter
L = distance
θ = angle of divergence

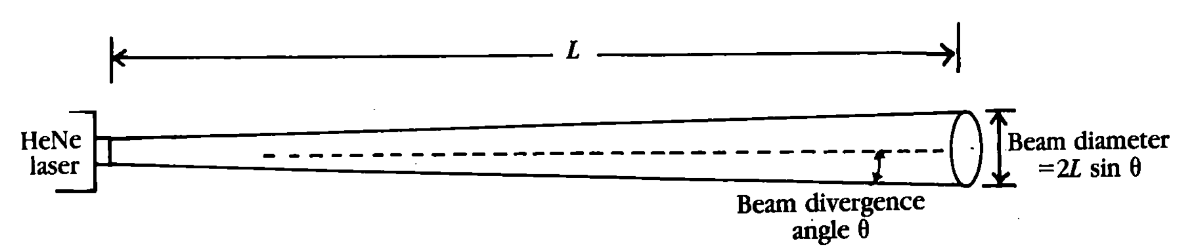

12-13 HeNe laser beam divergence.

EXAMPLE:

Using an HeNe laser with a beam divergence of 1.2 milliradians, calculate the diameter of the beam at 30 meters. First convert 1.2 milliradians into degrees:

$$1.2 \times 0.05729 = 0.06874$$

Then plug the information into the equation:

$$\begin{aligned} D &= 2(30) \times \sin 0.06874 \\ &= 0.071 \text{ meter} \end{aligned}$$

If the laser beam divergence is under 100 milliradians (0.1 radian or 6 degrees), you don't have to bother calculating the sin; the angle in radians is good enough to use directly:

$$\begin{aligned} D &= 2(30) \times 0.0012 \text{ radian} \\ &= 0.072 \text{ meter} \end{aligned}$$

SEMICONDUCTOR LASERS

Semiconductor lasers go by quite a few names: diode, injection, junction, and *junction diode* laser. The semiconductor laser has been around since 1962. Robert N. Hall of General Electric Research and Development made the first successful semiconductor laser. To understand the operation of this family of lasers, you need to know the basic properties of the semiconductor materials used in its construction.

Semiconductors

A semiconductor is a material whose properties lie between an electrical conductor and an insulator; hence, the name *semiconductor.* There are many semiconductor materials available today, but the most popular semiconductor material in use today is silicon, which can serve as a model. The atomic number of silicon is 14. See FIG. 12-14. A silicon atom has four electrons in its outermost shell. It requires four additional electrons, making a total of eight electrons to complete this outer shell. The number of electrons in the outermost shell determines the valence of the atom. For this reason, the outer shell is called the *valence shell,* and the electrons in the valence shell are appropriately called *valence electrons.*

Silicon can easily combine with other atoms in an effort to share electrons and fill its outershell. Pure silicon forms into a crystalline structure with four other neighboring silicon atoms (see FIG. 12-15). Each atom joins to another silicon atom by sharing an electron with the other. This shared electron bond is called a *covalent bond.* To keep things simple, FIGS. 12-15 to 12-17 show only the outer-shell electrons of the silicon atoms formed into a crystalline structure. Silicon in this pure form isn't very useful, but introducing a small amount of another compound into the silicon crystal structure dramatically changes the electrical characteristics of the material. The process is known as

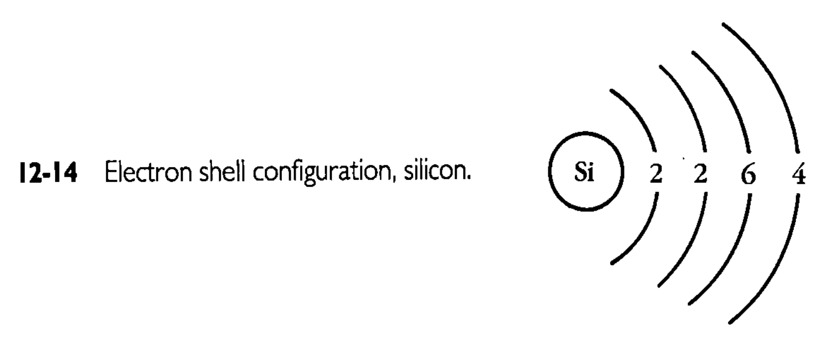

12-14 Electron shell configuration, silicon.

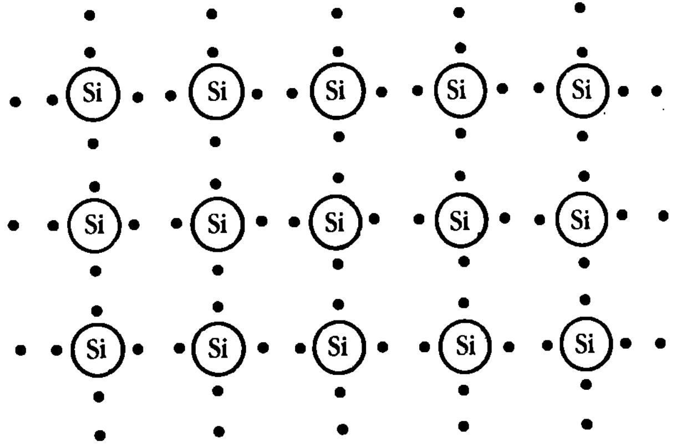

12-15 Crystal structure, pure silicon.

doping. Some typical compounds used for doping silicon are boron, arsenic, and phosphorus. The amount of doping material is on the order of 1 part in 108. Even at this small amount, the conductivity of the material increased by about a factor of 12.

P-type silicon A boron atom has only three electrons in its outer shell. By doping the silicon crystal with a small percentage of boron, the lattice forms with an electron missing from one of the covalent bonds (see FIG. 12-16). The vacant electron position is called a *positive hole*. It's possible for an electron from a neighboring atom to "jump" into this hole. In doing so, the electron creates a positive hole in the atom from which it leaves. Electrons and therefore electricity can flow through the crystal by jumping from hole to hole. From the outside, this electron jumping makes it appear as if the holes are also moving through the crystal silicon, just like bubbles move through water. Electrical conduction in this type of doped silicon is through the positive holes in the crystal lattice. It is therefore called p-type semiconductor (p stands for positive).

N-type silicon An atom of arsenic has five electrons in its outer shell. If the silicon is doped with a small percentage of arsenic, the crystal lattice formed will have an extra electron (see FIG. 12-17). The extra electron is mobile and can move freely through the lattice. This material is called an n-type semiconductor because electrical conduction through the lattice is achieved by the flow of the negatively charged electron (n stands for negative).

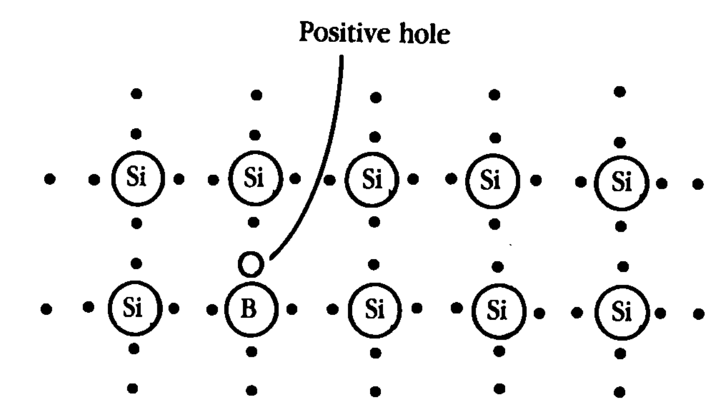

12-16 Crystal structure, p-type silicon.

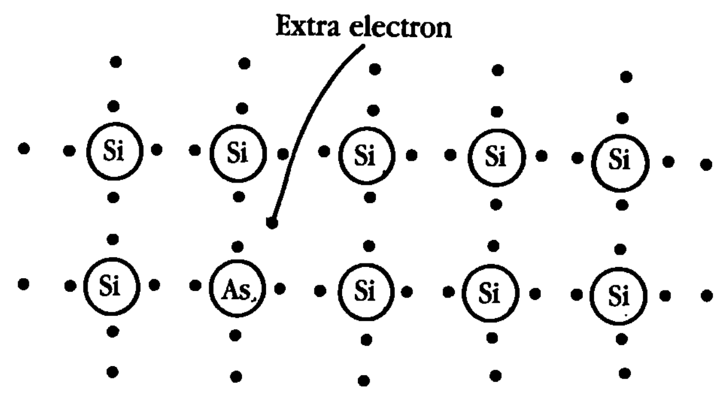

12-17 Crystal structure, n-type silicon.

Diodes

Both the n-type and p-type silicon conducts electricity in any direction, but the method is different. To reiterate, in the n-type semiconductor, the extra electrons are the current carriers, while in the p-type semiconductor, the positive holes are the current carriers.

If a piece of p-type and n-type semiconductor are joined together, the surfaces where the materials meet is called a pn junction. A pn junction is commonly known as a *diode*. A diode conducts electricity in only one direction. Figure 12-18 provides an explanation.

In FIG. 12-18A the diode is forward biased. The negative terminal of the battery repels the electrons in n-type material towards the pn junction. The positive terminal of the battery repels the holes in the p-type material towards the pn junction. The repelled charge meets at the pn junction. If the voltage is greater than 0.65 volts (for silicon), the electrons will have sufficient energy to jump through the junction and into the positive holes. The 0.65 volts represents the minimum amount of energy an electron requires to jump a silicon pn junction. Different semiconductor materials have different minimum threshold energies; for example, germanium only requires 0.035 volts. The process of the electrons crossing the pn junction and jumping into the positive holes is called *recombination*.

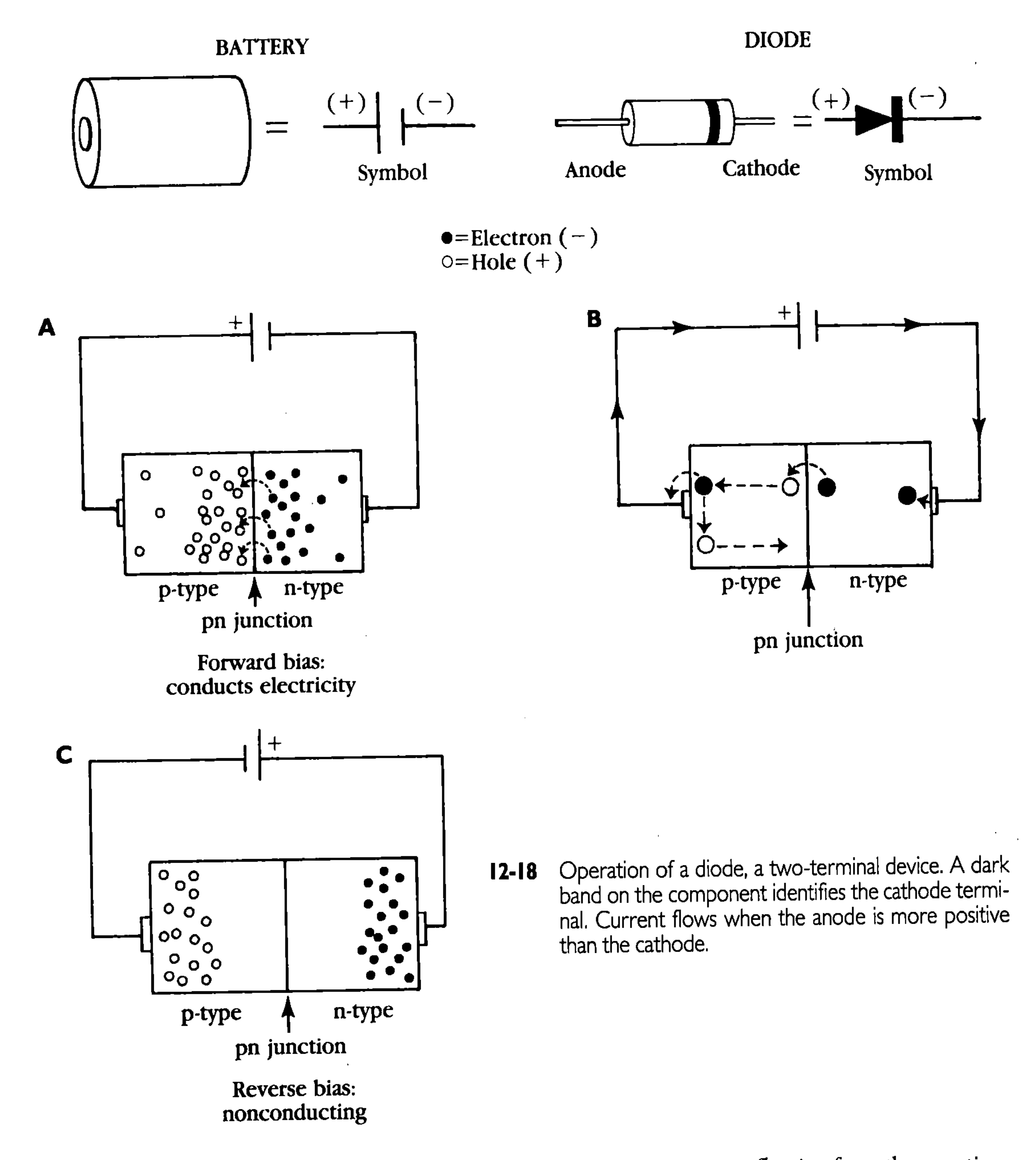

12-18 Operation of a diode, a two-terminal device. A dark band on the component identifies the cathode terminal. Current flows when the anode is more positive than the cathode.

This electron movement constitutes a current, flowing from the negative terminal of the battery through the n-type material, across the junction, through the p-type material, and back to the positive terminal of the battery.

Perhaps this would be easier to see and understand if you traced a single electron's journey through the diode. See FIG. 12-18B. As one electron jumps through the pn junction, an electron from the negative battery terminal enters the n-type material simultaneously. The electron that passed through the junction jumps into a positive hole and is attracted toward the positive terminal of the battery. When it reaches the terminal, it jumps out of the hole into the terminal where it completes its journey to the battery. When the electron jumps into the terminal, it leaves a positive hole that would quickly repel back to the pn junction where the process would be repeated. (In actuality, the hole/electron pair wouldn't travel. The electron could make numerous hole-to-hole jumps to the positive terminal. But the end results are the same.)

In FIG. 12-18C, the diode is reverse biased. Here the negative terminal of the battery attracts the positive holes to its side, while the positive terminal does the same for the electrons. There is no recombination at the pn junction, so there isn't any way for the current to flow through the diode. Therefore, no current flows.

Light-emitting diodes (LEDs) The pn junction of a diode can emit photons of light, depending on the composition of the p and n materials. When an LED emits light, it does so with greater efficiency than most other light sources. Look at FIG. 12-19. When the electron jumps through the pn junction, a photon

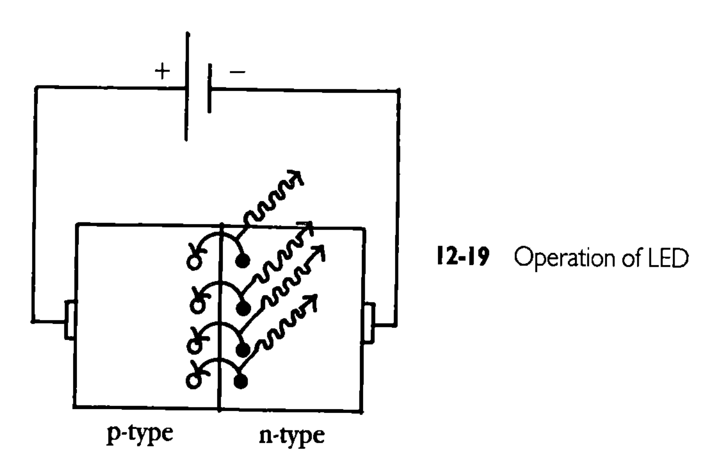

12-19 Operation of LED

of light is emitted as it falls into the positive hole, quite similar to the atomic energy transitions in the atom. The energy required for the electron to jump through the pn junction is called the *bandgap.* This energy, expressed in eV, can be used to calculate the wavelength of the emitted photon, and it is the same relationship as used for atomic transitions;

$$1 \text{ eV} = 1242 \text{ nm}$$

In the example used to show the operation of the diode, it was stated that the threshold energy required of the electron is + 0.65 eV. You calculate this energy to the wavelength of the photon emitted as follows.

$$\frac{1242 \text{ nm}}{0.65 \text{ eV}} = 1.91\ \mu\text{m}$$

This emitted photon is in the far infrared portion of the spectrum and most likely warms the crystalline structure. Visible light photons have an energy value in the range of 1 eV to 3 eV.

Because all the electrons jumping through the pn junction typically require the same energy, the LED emits a narrow band of wavelengths (see FIG. 12-20). The emitted photon's energy doesn't exactly match the bandgap energy (the energy required to cross the pn junction), but it is a close approximation. In addition, the voltage drop across the diode (expressed in eV) is a close approximation of the bandgap energy.

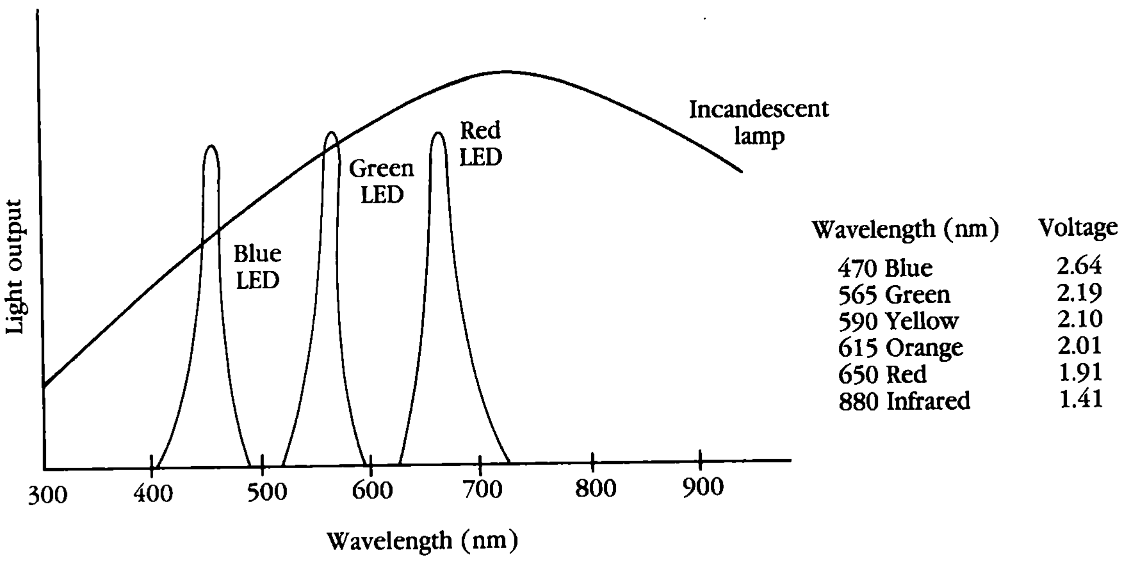

12-20 Wavelength of various LEDs.

Diode laser

Lately there has been a great deal of progress in laser diode technology. The impetus for the research and development can be laid at the door of many viable commercial uses for this product: CD players, laser printers, optical information disks for computers called WORM (*w*rite *o*nce *r*ead *m*any) drives, bar-code readers, pointers, gun sights, and of course fiberoptic telecommunications. The holography community stands to benefit from this laser diode development. As the technology continues to develop, the cost will drop as the quality improves.

A typical laser diode is illustrated in FIG. 12-21. The laser diode and the LED are closely related. Light is generated from the recombination of electron/hole pairs at the pn junction. The optical cavity of the laser diode is short, typically about 500 micrometers long. The mirrors are cleaved faces of the semiconductor material. Typically, the mirrors reflect only 30 percent of the incident light back into the crystal. The laser functions even with this low reflection ratio due to the high gain of the semiconductor material. The profile of the laser beam generated is usually elliptical. Because of the short optical cavity length I would have thought that the laser diode would be very coherent; however, this is not

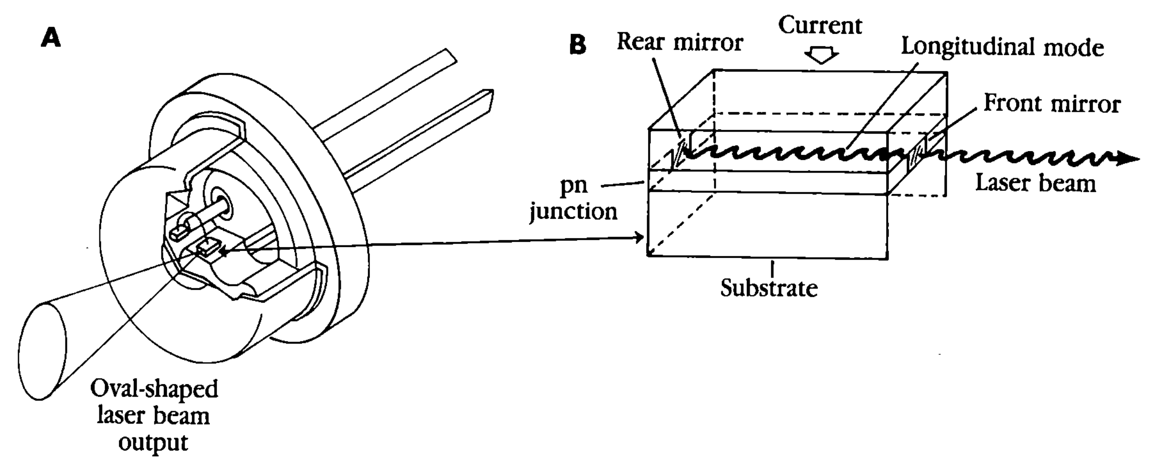

12-21 (A) Cutaway diagram of laser diode and (B) detail of laser diode chip.

the case. The optical cavity can also support longitudinal wavelengths as the HeNe laser. This limits the coherency of the laser diode, making most of them unsuitable for holography. But there are ways to improve the coherency of the diode. Laser diodes have an estimated life of 10,000 hours. Diodes exist that can operate in either pulse or continuous wave (cw) mode.

Homostructure and heterostructure The first laser diodes used the same semiconductor base material for both the p-region and n-region. These are called *homostructure* or *homojunction* lasers.

As development on laser diodes continued, researchers tried different base semiconductor materials within a single diode. These diodes are *heterostructure* or *heterojunction*. The advantage found by the researchers in using different materials was that the laser became more confined in the pn junction, making the diode more efficient. This happened because the refraction index of the semiconductor materials were different (see chapter 11). Figure 12-22 illustrates a homostructure and a heterostructure diode.

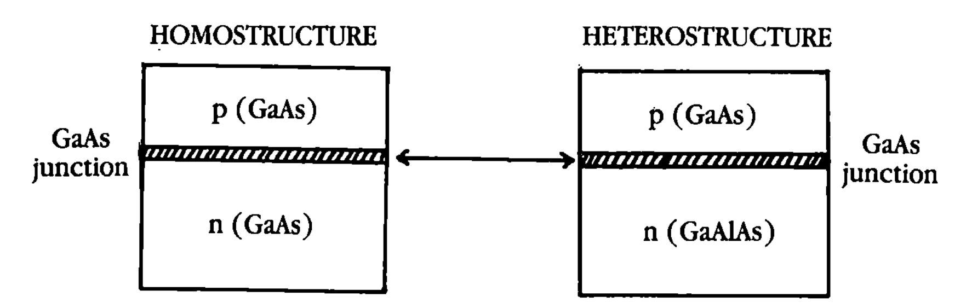

12-22 Cross section homostructure and heterostructure diodes.

Strip-geometry diode lasers If you look at the illustrations for the homostructure and heterostructure diodes, you can see that if these diodes were made to lase, they would produce a beam across the entire width of the diode. To improve the quality of the emitted beam, the width of the active junction is reduced to a strip. Figure 12-23 illustrates a cross section of a strip-geometry laser diode. The insulators on top concentrate the current flow into the strip. Only the current flowing in the strip has sufficient power to support laser activity. Various types of semiconductor materials are used to help define the beam. Again, this is through the different refractive indexes of the semiconductor materials.

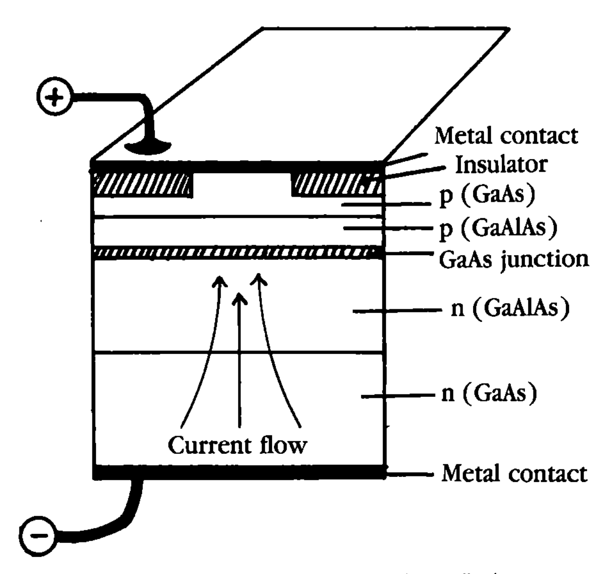

12-23 Strip-geometry laser diode.

Distributed feedback As state previously, the optical cavity of the diode laser supports longitudinal modes, which ruin the coherency of the emitted laser beam. One method employed to limit the longitudinal modes and thereby increase the coherency is the distributed feedback laser, shown in FIG. 12-24. This technique uses a diffraction grating that scatters light back into the active layer. The feedback of light causes interference effects that allow laser oscillation to occur only at the wavelength that is reinforced by the scattering effects. The grating spacing determines the particular wavelength that is reinforced. This technique is employed in the newer visible laser diode and limits the bandwidth of these diodes to 0.2 nm.

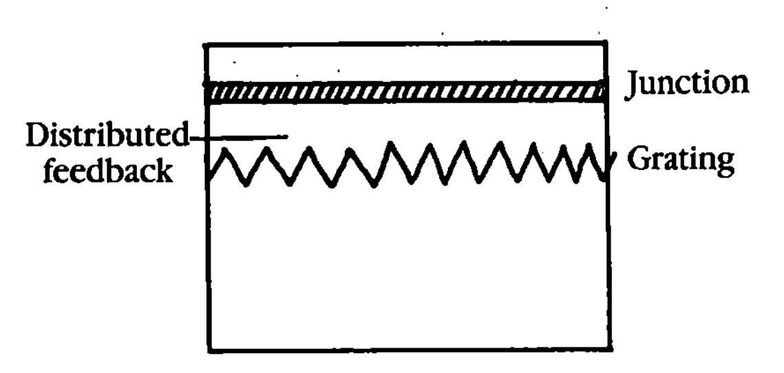

12-24 Distributed feedback laser diode.

Beam divergence Because of the small size and shape of the active layer in the laser diode, the beam divergence is great. The overall shape of the emitted beam is an oval. Fortunately, this can be corrected with external optics.

Impact on holography

There is currently one company in the United States that claims to produce a laser diode suitable for use in holography: Holo-Spectra (7742-B Gloria Ave., Van Nuys, CA 91406 (818) 994-9577). Holo-Spectra uses a diode manufactured by the Japanese company Toshiba in their HSD-4 laser diode assembly. The wavelength is 670 nm, estimated life of the unit is 100,000 hours, power output is 3.4 milliwatts, beam divergence is 1.2 milliradians (with external optics), and the cost of the unit is $575 (as of January, 1990). The coherency length of the diode isn't known, but it has been used by a few American holographers to produce holograms.

As laser diodes develop, the cost should drop and the quality should improve. Also, shorter wavelengths would be a tremendous boon for color holography.

660 nm laser diode

More recently, Toshiba Corporation announced a 660 nm laser diode. As of January, 1990, this is the world's shortest-wavelength laser diode produced. The diode uses a light-emitting layer of indium gallium aluminum phosphide. Because of its shorter wavelength, the laser's perceived brightness is double that of its 670 nm predecessor. In the United States, samples are available from Toshiba America Electronics Corporation, Irvine, CA. The laser diode's part number is TOLD9220, and the cost is approximately $200. No information is available concerning coherency length; I therefore assume it's quite low.

OTHER TYPES OF LASERS

There are various types of lasers available today. Some of them are suitable for different aspects of holography. Following is a brief description on other types of lasers.

Helium-cadmium (HeCd) The helium-cadmium laser produces laser light at a wavelength of 442 nm. The HeCd laser can be used (in conjunction with other lasers) to produce color holograms. Because of its shorter wavelength, it can be utilized to produce holograms on a variety of nonsilver emulsions that are relatively insensitive to the HeNe red light. The coherency length is somewhat shorter than an HeNe laser, and tube life is generally around 5,000 hours.

Argon ion (Ar^+) Argon lasers can produce laser light at a number of different wavelengths. The most common wavelengths used on the argon laser are 488 nm (blue) and 514.5 (green). Power output is typically 0.5 to 1 watt. Because of its high power and short frequency, this laser is employed in professional transmission holography. The argon laser is not without disadvantages. Argon lasers operate at high temperatures so forced air or water cooling is necessary. The high temperature also creates a wide Doppler broadening in addition to numerous longitudinal modes. This laser must be used with an etalon or some other method to increase the coherency of the laser.

Solid-state lasers Typical of the ruby laser using different host materials in place of the ruby crystal.

Dye lasers Composed of organic dyes usually dissolved in a liquid solvent. The dyes typically have a broadband of wavelengths making the laser tunable throughout the band. Various dyes exist that cover different parts of the spectrum. These lasers can be operated in pulse or continuous mode but commonly are operated similarly to solid-state lasers.

There are a host of other laser types: metal vapor, chemical, free electron, but I feel I have covered what are the most important lasers that relate to holography today.

LASER CONTROVERSY

Older texts on lasers credit Schawlow and Townes with the invention of the laser because of an approved patent application submitted by these scientists in 1958. However, a graduate student named Gordon Gould claimed to have invented the laser in November, 1957. Over the years, Gould mounted an attack against the United States patent office. After years of struggle and about 6 million dollars in legal cost, the U.S. Patent Office reversed the earlier patent given to Schawlow and Townes and gave it to Gordon Gould. Gould now enjoys the royalties from companies who manufacture lasers. The manufacturing companies are justifiably angry after paying royalties on the first patent for 17 years, and now they have to contend with another patent. For Gould, the delay is a boon because the patent is worth much more money now than at the birth of laser technology because there are many companies manufacturing and using lasers today.

Although this changes the inventor of the laser, the first person to successfully construct a laser remains to be Theodore Maiman.

Chapter 13

Diffraction model

Previous chapters explained that making a hologram requires recording the interference pattern generated between two wavefronts. This interference pattern recorded in holographic emulsion is actually a diffraction grating. To see this more clearly, recall the double-slit experiment of Thomas Young. Each slit behaved as a separate light source. The interference pattern created was displayed on the screen. The two light sources in this demonstration were adjacent to one another along the same axis.

Figure 13-1 illustrates two off-axis waves generating an interference pattern. Because both beams are incident upon the plate from the same side, the interference pattern created is analogous to a transmission type of hologram. Since we are using two plane wavefronts incident on an angle, this creates what is commonly known as a *sine grating*. The angle of incidence between the two wavefronts determines the spatial frequency (spacing distance) of the interference pattern in accordance with the following equation:

$$f = \sin\theta/\lambda$$

where λ is the wavelength of light.

For all the following discussions, assume an HeNe light source (632.8 nm) for all beams unless otherwise noted. In FIG. 13-2, the recorded interference pattern is a hologram (transmission type) of wavefront B. After the emulsion is developed, by illuminating the emulsion with just the reference beam (labeled A) the light is diffracted. The first-order diffraction from the holographic sine grating propagates on either side of the transmitted beam. One diffracted wavefront behaves as if it originated from wavefront B used in making the hologram. In the illustration, this beam is traced backwards using dotted lines. This diffracted beam is labeled "virtual beam B" in FIG. 13-2. This is a brief overview of how holograms are recorded and reconstructed, but let's continue by using an object to clarify things further.

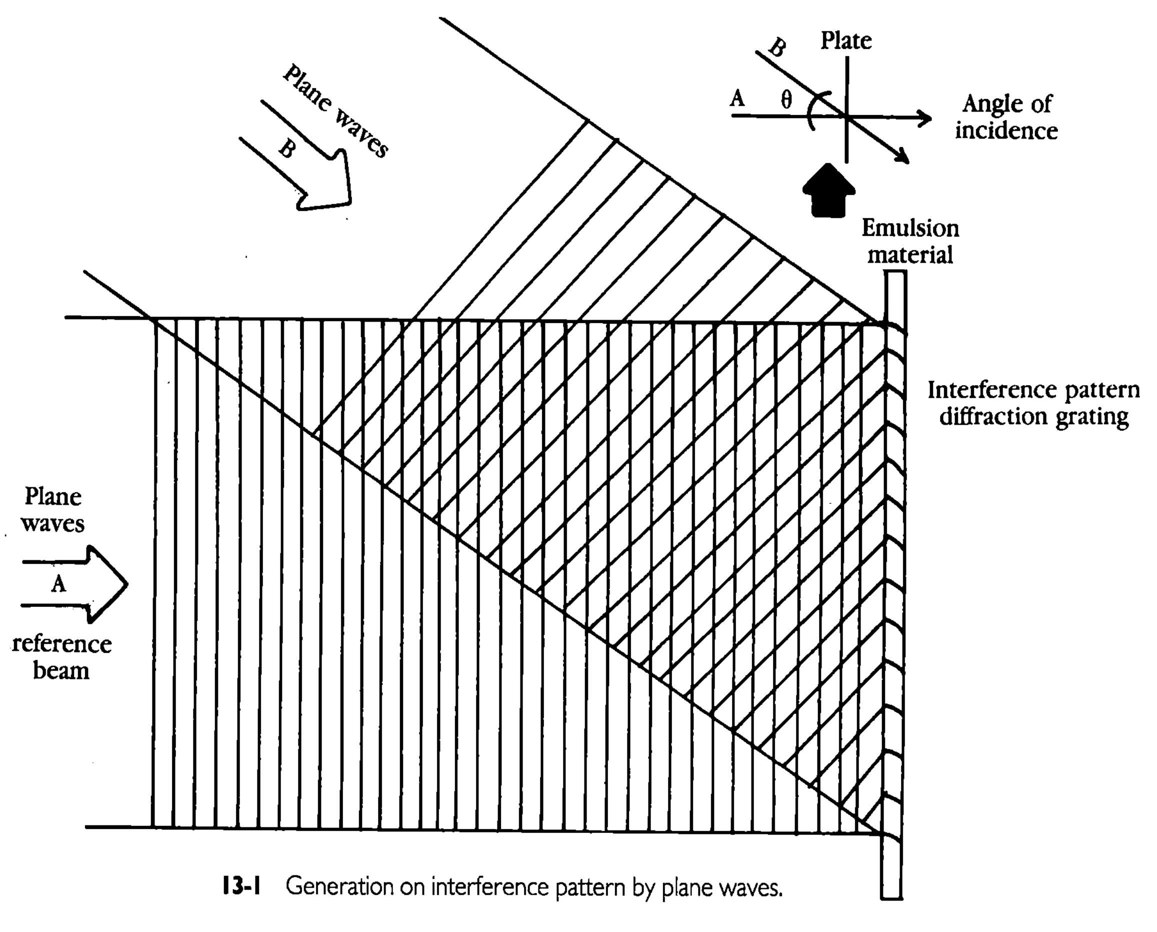

13-1 Generation on interference pattern by plane waves.

GABOR ZONE PLATE

To keep the illustration simple, let's begin by analyzing the recording and reconstruction of a single point of an object being holographed. Using a single point is a good idea for two reasons. One is that it keeps the illustration simple, and secondly, any object can be considered as a large number of adjacent points that are all being holographed at the same time. The size of the points, in practical consideration, depends on the wavelength of light and the resolution of the film being used.

The spherical wavefront emanating from point A in FIG. 13-3 represents the reflected light off of the object from the object beam. The reference beam is shown incident on the plate at an angle. Again, to keep the drawing simple, the reference beam is shown at an angle. To generate a *zone plate,* the reference beam should be incident along the same axis as point A. The interference pattern created by these two wavefronts are a series of concentric circles, as illustrated in the front view of the plate; this is the zone plate and its properties have been studied for many years. To be more specific, the plate created is a Gabor zone plate, and it operates similarly to the classical or Fresnel zone plate.

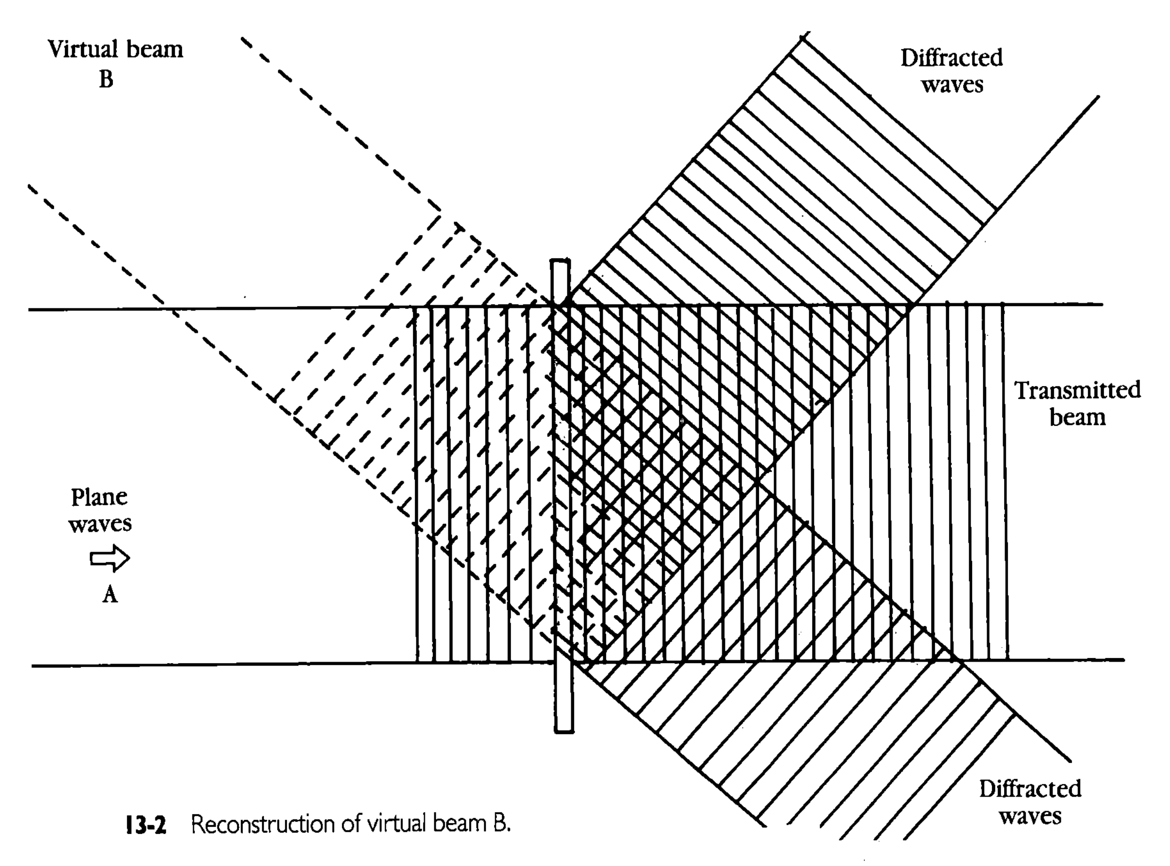

13-2 Reconstruction of virtual beam B.

Figure 13-4 illustrates both types of zone plates. The classical zone plates have abrupt changes from transmission to opaque regions, while the Gabor zone plate changes from transmissive to opaque as a sine function. Light incident upon the zone plate is diffracted by the concentric circles. The concentric rings are arranged so that one portion of the diffracted light converges to a single point (focus length) and gives a real image of point A. The diffracted light that diverges, if traced back as indicated by the dotted lines, corresponds to the virtual image of point A.

Figure 13-5 is an overview of the entire image reconstruction. Imagine that a hologram is composed of a large number of Gabor zone plates, each of which reconstructs a single point of a holographed object.

So far, red HeNe light has been used to create and reconstruct the image. Suppose you made a transmission hologram and illuminated it with white light. Because white light is composed of many different frequencies, each wavelength will diffract to a slightly different angle, the results of which would be a fuzzy, multicolored image of the object. This is why transmission holograms need a monochromatic light source for viewing. Transmission holograms can be viewed successfully with white light using a narrow band interference filter in front of the light source. The narrow band filter blocks most of the wavelengths and allows just a small portion of the wavelengths to pass through.

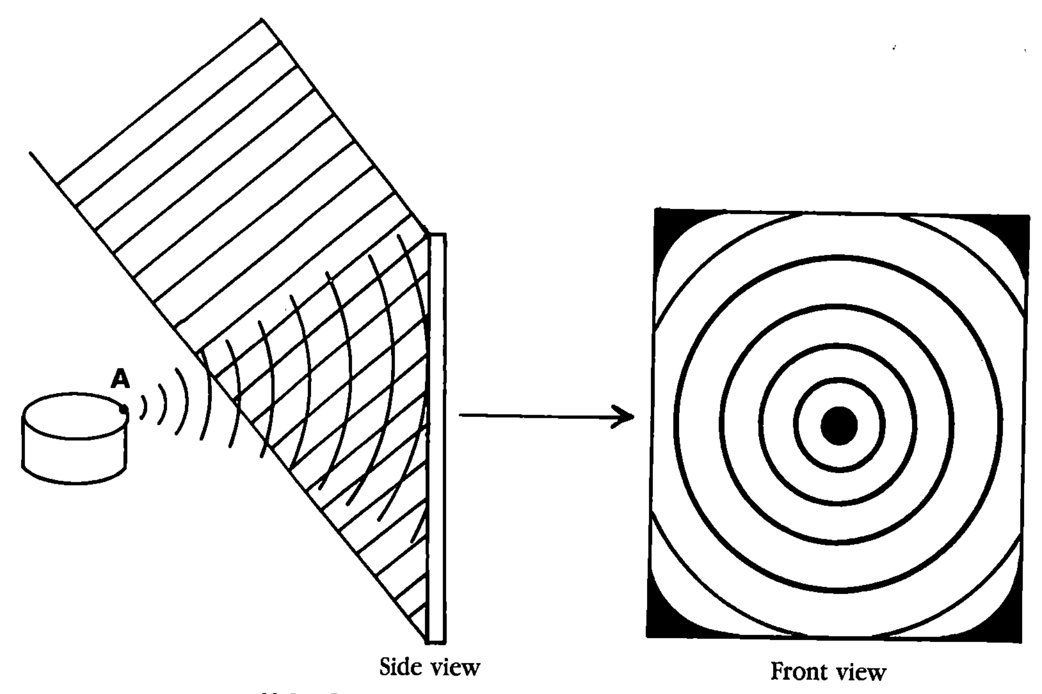

13-3 Generation of point hologram (emulsion zone plate).

It might appear that transmission holograms are less useful than white light reflection holograms, but this is not the case. Good transmission holograms can be used to create full-color, achromatic (black and white), rainbow, and reflection copy holograms.

BRAGG DIFFRACTION

The discussions of the transmission holograms have inferred that the emulsion thickness is infinitesimally thin and diffraction describes the reconstruction of the real and virtual images. The diffraction model needs to be supplemented with Bragg diffraction planes to clear up an apparent discrepancy in the diffraction model. The discrepancy is that according to the diffraction model, both the real and virtual images should be of equal intensity. However, this is not the case; usually when observing the virtual image, the real image is quite dim, and vice versa. To observe these images, you must flip the hologram around to properly view one image or the other.

The reason you cannot rely on just the diffraction model is that the emulsion material is several light wavelengths thick. The fringe pattern is created throughout the thickness of the emulsion material. Figure 13-6 illustrates the fringes in a transmission-type hologram, which leads to the Bragg condition. In order to achieve a bright image, the intensity of the diffracted beam needs to be high. The wavefronts diffracted by the fringes must be in phase with one another (constructive interference). Looking at FIG. 13-6, notice that reconstruction only occurs when the incident light is at the proper angle

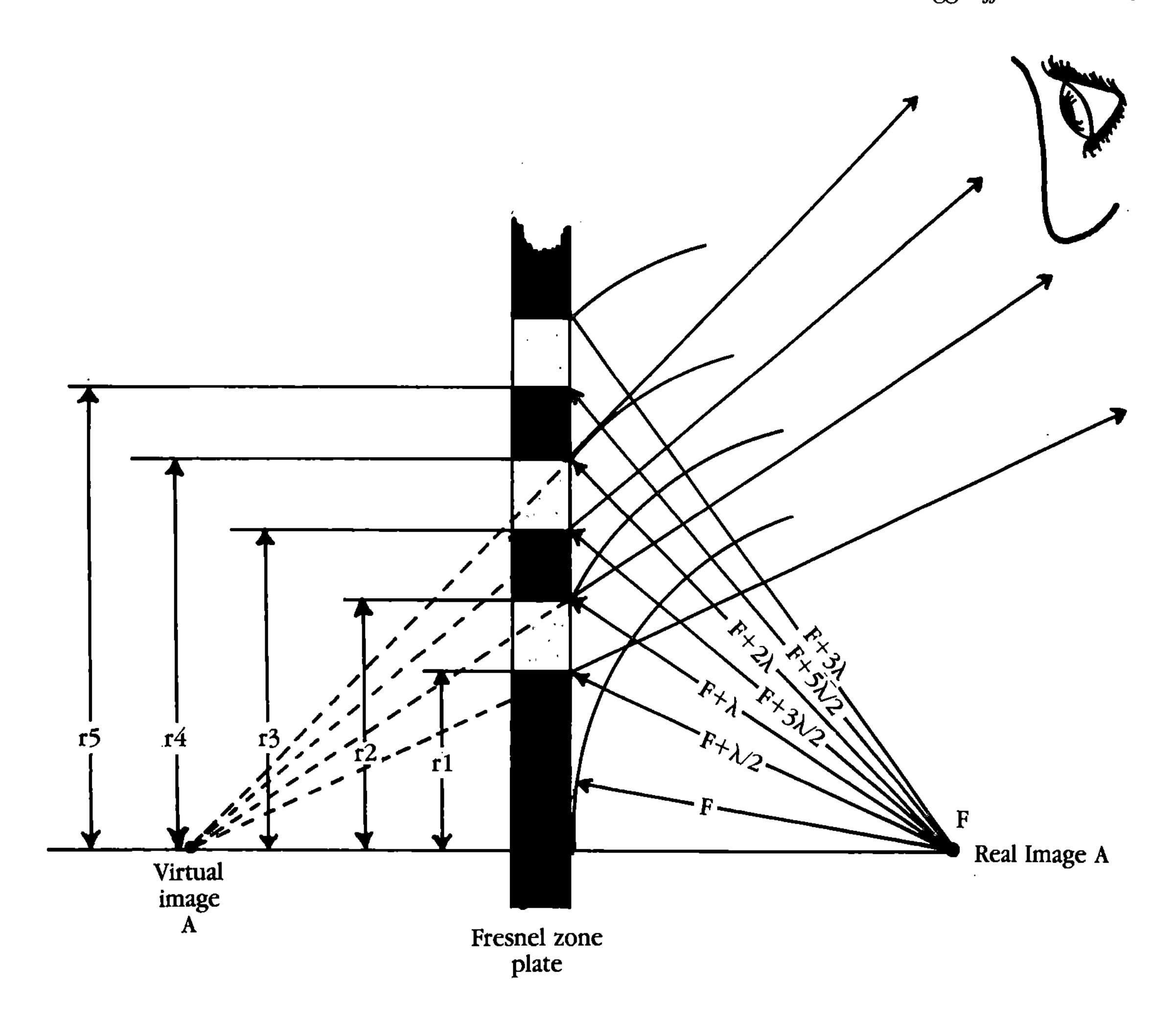

Gabor zone plate

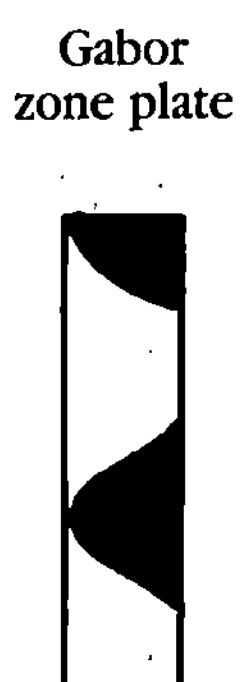

13-4 Operation of classical zone plate.

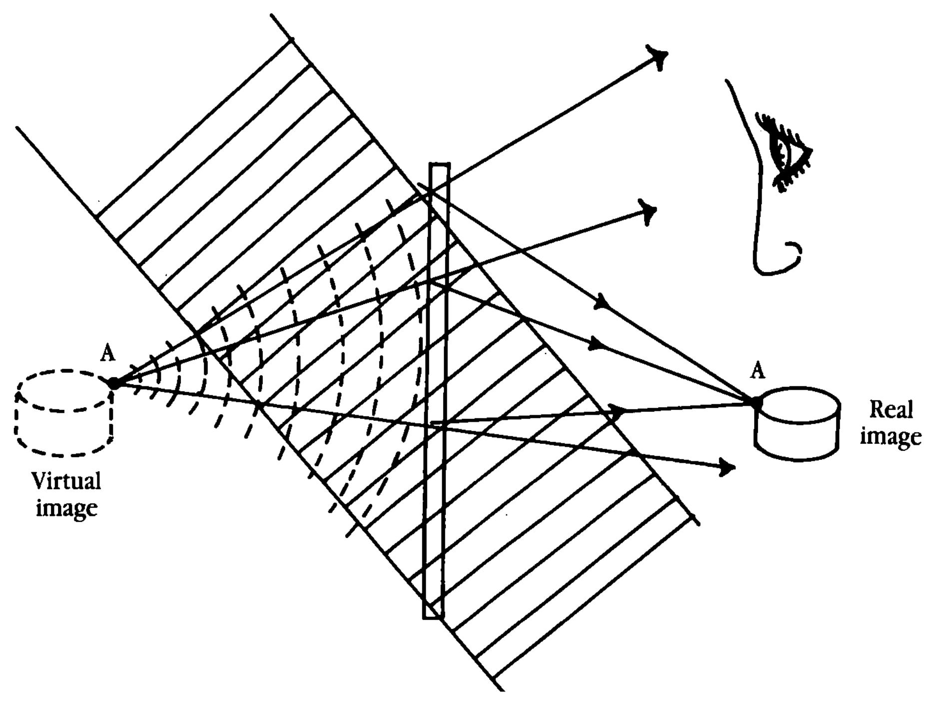

13-5 Overview of image reconstruction showing point hologram.

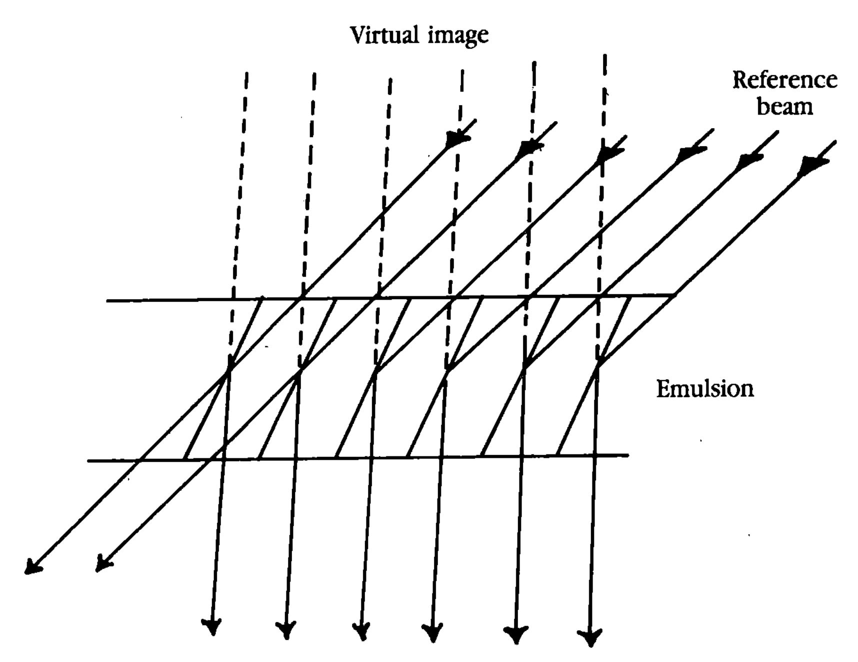

13-6 Operation of reference beam through Bragg diffraction planes (transmission hologram).

(determined by the angle of the reference beam when creating the hologram). If the incident light comes in too steep, the light will pass through the fringes without being diffracted. On the other hand, if the angle comes in too wide, the light is reflected and diffracted out of phase and no image can be reconstructed.

In FIG. 13-6, I take the liberty of showing just the light being reflected in phase off of the fringe pattern. In practice, when the Bragg condition is met for the virtual image, it is not met for the real image and vice versa.

Bragg diffraction in reflection holograms

When making reflection holograms, the interference patterns lie more parallel to the emulsion surface due to having the reference beam and object beam incident on the plate from opposite sides. The fringes created are typically about one-half wavelength apart (see FIG. 13-7). Reflection holograms are viewable in white light. Observe the behavior of light as it passes through the interference pattern (Bragg planes). When white light is incident on the emulsion (at the proper angle), only light of the proper wavelength will be reflected and diffracted in phase and reconstruct the image. All other wavelengths either pass through or are canceled through destructive interference. Now, since the fringe spacing is proportional to the laser light that created it,

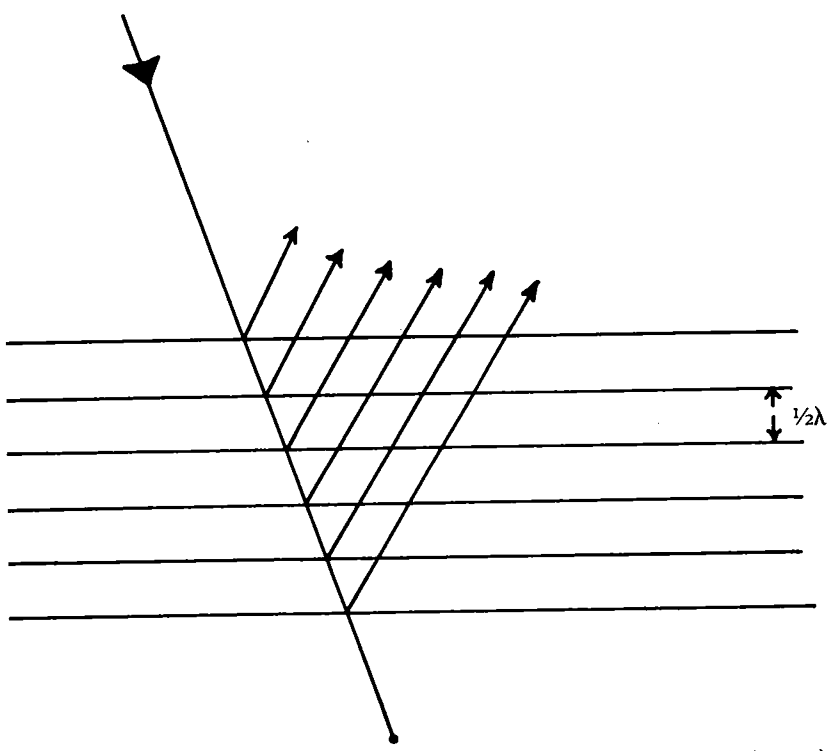

13-7 Operation of reconstruction beam through Bragg planes (reflection hologram).

the Bragg condition is met only for the red portion of the white light. This explains why the image usually reconstructs in red, the same color it was holographed in.

By increasing the angle of incidence of the white light, the image color shifts towards the blue because as the angle of incident light increases, the spacing between the fringes decreases. Now the Bragg condition is met for shorter wavelengths and the image reconstructs at a color of higher frequency.

PSEUDOCOLOR

It becomes a simple matter to tie in pseudocolor holograms. Recall that one method of generating pseudocolor holograms began by swelling the emulsion with TEA solution. When using such a treated plate, the interference pattern recorded would still be approximately one-half wavelength apart. However, during development, the swelling agent washes away, allowing the emulsion to return to its original size, which subsequently shrinks the spacing of the interference pattern. The Bragg condition for the hologram is now met at a shorter wavelength than the laser light used to produce the hologram, which explains how the pseudocolor reflection holograms are generated. Also note that during some development processes, the emulsion material shrinks, which also causes the hologram to play back at higher frequencies (shorter wavelengths).

FOURIER TRANSFORMS

Sooner or later, if you stay into holography you'll hear mention of Fourier optics, diffraction, holograms, computer-generated holograms, transforms and so on. Although the mathematics behind Fourier transforms is beyond the scope of this book, I can give a general overview on some aspects of Fourier optics.

Basically, Fourier optics give a good mathematical description of the behavior of light through a lens and diffraction. In regard to diffraction, a Fourier analysis adequately describes far-field diffraction, or *Fraunhofer diffraction* patterns. A lens can take an approaching wavefront of incoherent light and transform this information into a new wavefront that creates a viewable image. Fourier optics are named after Joseph Fourier. He was a scientist who worked for Napoleon. Fourier developed his mathematical analysis by studying the propagation of heat in a metal ring. He discovered that heat propagated through the ring in a sinusoidal manner. In addition, any wave could be considered the sum of an infinite number of sine waves. The results of his method found far-reaching applications in electronics, communication technology, and most recently in optics and holography.

There are quite a few books that deal exclusively with Fourier transforms for those of you who wish to pursue this aspect of holography. Fourier optics are being brought forth in a few cutting-edge technologies such as pattern recognition by computers, mass storage devices, and optical elements.

COMPUTER-GENERATED HOLOGRAMS

Computer-generated holograms are becoming a hot topic. It is a relatively simple-to-understand process. Most computer-generated holograms are of two-dimensional objects, typically a line drawing or printed text on a transparency. A computer calculates the interference pattern that is generated when a monochromatic light source (laser light) is incident on the transparency. All of this happens inside the computer, so the object doesn't necessarily exist physically. Usually, the picture is held as binary numbers in a data array. The program slowly calculates, point by point, the interference pattern. The calculated interference pattern is printed or plotted onto paper. The pattern is then photographed from the paper to a transparency and reduced. The amount of reduction depends on various parameters such as size of the plot paper to photocopy and the resolution of the transparency film. These parameters also determine the field of view and apparent size of the image. The transparency, when illuminated with laser light, reconstructs the original image.

Some of the programs used to create these interference patterns use a FFT (*f*ast *F*ourier *t*ransform) program. Others I have seen take a more classical approach and use standard diffraction equations. Both generate viewable holograms. Many hope computer-generated holograms become useful for generating HOEs (*h*olographic *o*ptical *e*lements).

SHORT HISTORY OF HOLOGRAPHY

Holography takes root in 1947. Dennis Gabor was working on a method to increase the resolution of the electron microscope. He had hoped to achieve this by making a hologram of a specimen using coherent x-rays and reconstruct the image using light. An image so made and reconstructed would be greatly magnified 3-D image.

Although this particular aim was never achieved, Gabor laid the foundation of holography. Gabor was able to demonstrate his principle of holography by making a hologram of a two-dimensional transparency. The depth limitation of the hologram at this time was due to the fact that a suitable coherent light source was not available. The best Gabor could find was a well-filtered mercury arc lamp that gave him a coherency length of 1 millimeter. Gabor coined the term *hologram.* The roots of the word are *holos,* meaning *whole* or *entire* and *gram,* referring to something written or drawn. The word *holograph* had already been taken by the literary community to mean something written in an author's own hand. But today its meaning more often refers to three-dimensional holography. In 1971, Dennis Gabor was awarded the Nobel Prize in physics for his discovery of holography.

In 1962, a soviet scientist in the the USSR, Yuri Denisyuk, developed the white light reflection hologram. Denisyuk developed the single-beam white light hologram (described earlier in this book). Because the process showed similarities to the Lippman process, Denisyuk called this type of hologram a Lippman Hologram. Today, this type of hologram is called a Denisyuk, Lippman, white light reflection, or reflection hologram.

In 1963, Emmett Leith and Juris Upatnieks, working at the University of Michigan, began producing holograms with a new coherent light source: the laser. The laser provided them with a much greater coherency length than had previously been possible. This allowed them to create holograms of great depth. They had also solved a serious problem encountered by Gabor who used an on-axis beam to create holograms. Leith and Upatnieks used an off-axis reference beam that separated the reconstructed real and virtual images and allowed for much better viewing. The type of holograms produced were exclusively transmission holograms.

In subsequent developments, in 1965, holographic interferometry papers were published by Robert Powell and Carl Stetson. In 1968, Steve Benton produced a transmission hologram that could be viewed in white light. And in 1947, Michael Foster created a method of mass-producing holograms cheaply.

Chapter 14

Building a laser

The first question is whether you should buy or build. New HeNe laser power supplies cost $150 and up. Surplus power supplies are sometimes available from companies like Meredith Instruments MWK Co. and Allegro (see "Sources"). These used, or *surplus,* power supplies cost about $75. Building your own power supply costs approximately $40.

The power supply circuit described here is quite simple. If you decide to build your own power supply, instructions follow. If you decide instead to purchase a surplus power supply, you can skip the construction information, but please read all the other sections on taking precautions and connecting and adjusting the power supply with the laser tube. If you are buying a surplus power supply, read the section "Surplus Power Supplies" near the end of this chapter.

A FEW WORDS OF CAUTION

All HeNe laser power supplies are high-voltage devices. Some people become careless with these power supplies because the current is relatively low (typically 4 to 7 milliamps). Do not become careless! You can still receive a nasty shock, and the resulting reflex muscle contraction could cause another injury. Be careful of all live wires. Make sure that all wires and components are properly insulated. Only work on the power supply when it is off, except for final adjustments with the laser tube. The power supply uses high-voltage capacitors that can hold a charge after the supply is turned off. When working on the power supply after it's been on, temporarily short the output leads of the power supply to drain any charge left in the capacitors.

POWER SUPPLY

The power supply for the laser is simple, but a little bulky. The reason for the bulk is that the circuit uses a 12-volt car ignition coil. This coil serves a dual purpose: as a high-voltage transformer and a trigger transformer. This simplifies the circuit design and keeps the cost down. The power supply output is adjustable, so it can be used for any HeNe laser tubes with output powers ranging from 0.5 milliwatt to 5 milliwatts.

Power requirement for an HeNe laser

HeNe laser tubes require a high-voltage power supply between 1,200 and 3,000 volts. The current is quite low, typically around 3 to 7 milliamps. These variables depend upon which tube you are powering. For instance, a 4-milliwatt laser requires 2,450 volts at 6.5 milliamps. A 1.5-milliwatt laser tube requires 1,350 volts at 5 milliamps. Your tube power supply requirements might vary depending upon the tube manufacturer. When purchasing a tube, new or used, this power supply information should be provided.

The trigger transformer is used to initially start the tube. It provides a momentary pulse of 8 to 10 kilovolts. This isn't a problem for the circuit, because the ignition coil doubles as an automatic firing trigger transformer to start the tube. Once the tube fires, the coil stops firing the trigger pulse and operates as a high-voltage transformer, keeping the HeNe tube lasing.

Construction

Figure 14-1 is the schematic for the power supply. All the parts are available from Images Company or any Radio-Shack store (see parts list in TABLE 14-1).

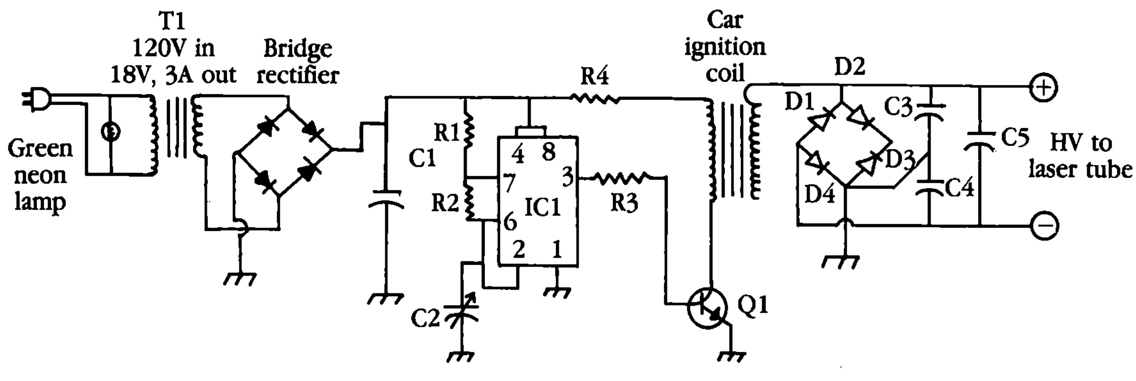

14-1 Schematic of laser power supply.

The circuit is simple, so you can use point-to-point wiring.

Practically any 12-volt car ignition coil can be used. It isn't necessary to purchase a new coil. A used coil purchased from a local scrap yard can work just as well and cost about one-fourth of the price. Most ignition coils are three-terminal devices. The two binding posts are where the coil connects to the circuit. The high-voltage terminal is the small tubelike projection in the center. Solder a wire to the metal inside the projection to create the high-voltage lead that connects to the high-voltage rectifying section.

The heatsink for the 3055 transistor is rather large and is important, so don't use a small heatsink. This circuit needs to run for hours at a time, and you don't want to have an overheated transistor breaking down in the middle of a holography setup.

Table 14-1 Parts List for Laser Power Supply

T1	120 Vac-to-18 Vdc, 3A transformer
C1	1000 μF, 35 V
C2	92–420 trimmer
C3, C4	0.01 μF, 6 kV
C5	0.05 μF, 5 kV
R1	470 kΩ
R2	2.2 MΩ
R3	220 Ω, ½ watt
R4	1 Ω, 10 watt
IC1	555 timer
D1–D4	10 kV diodes
Q1	3055 transistor
	Bridge rectifier
	Green neon lamp
	Plugs

The integrated circuit chip is a 555 timer. Radio-Shack sells two versions of this chip; a standard chip and a CMOS version. *Do not buy the CMOS version.* It doesn't handle enough power to drive the circuit. Use only the standard 555 timer.

Connecting the power supply

Always have a load connected to the output of the high-voltage power supply circuit when testing. If you don't, the high voltage produced could arc across different connections on the circuit board. The load can be the laser tube ballast resistor (50K to 250 kilohms), with or without the laser tube.

The helium-neon laser tubes requires a ballast resistor to limit the current flowing through the tube. Again, when purchasing the laser tube, new or used, the ballast resistor required is either supplied or its value is given. Never operate an HeNe tube without the ballast resistor, as this is sure to destroy it. Most ballast resistors range between 50 kilohms and 250 kilohms, with a 3- to 5-watt capacity. Figure 14-6 is a general drawing of connecting the power supply to the tube. The ballast resistor generally connects on the anode (+) side of the laser tube, with no more than a few inches of wire between the anode terminal of the laser and the ballast resistor.

The HeNe tube has a positive terminal called the *anode* and a negative terminal called the *cathode.* These terminals are connected to the positive and negative terminals respectively of the high-voltage power supply. Identifying the positive and negative terminals on the laser tube might not at once be obvious. Figure 14-2 shows a basic HeNe laser tube. Sometimes, the positive terminal is marked with an "A" (for anode), "+," or a small red dot. The negative terminal, the cathode, is sometimes marked with a "C" (for cathode) or "K." The cathode can also be identified by a small tube on one end that is used to fill it with gas.

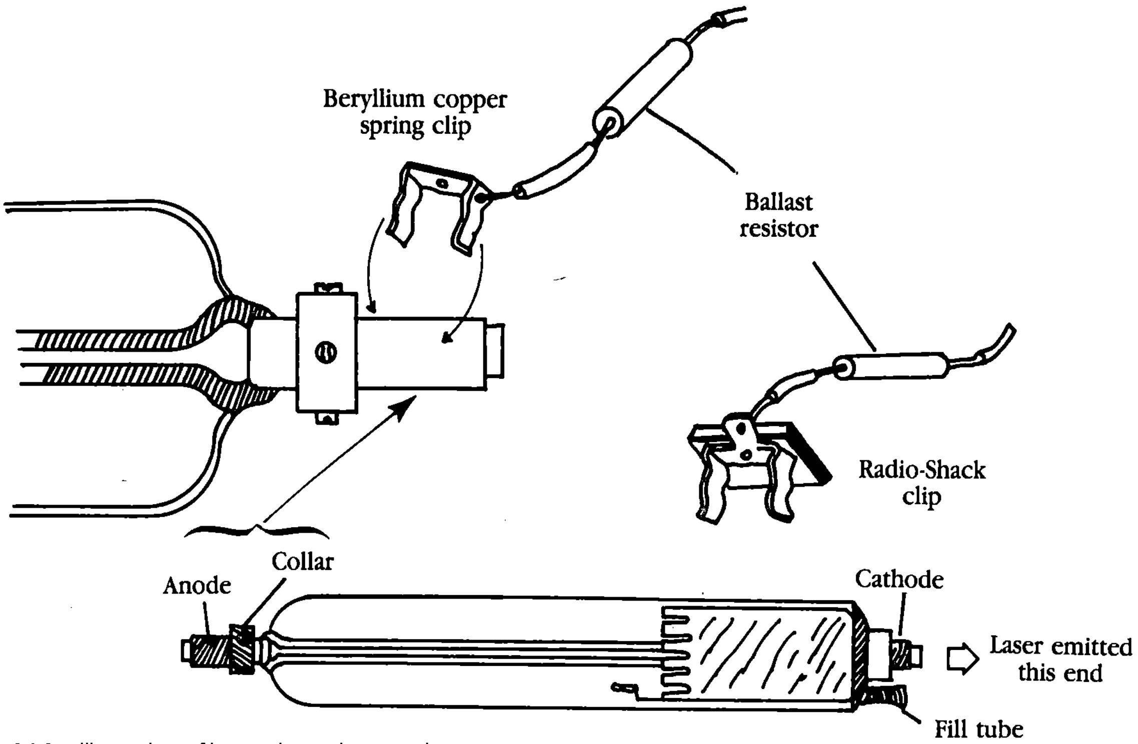

14-2 Illustration of laser tube and connections.

Making the laser tube connections

Most tubes manufactured today are hard sealed. The metal terminals on the end are also the mirror mounts. This makes it a bad idea to solder wires directly to the terminals. Some companies offer beryllium copper spring clips that clip onto the terminal. If they are available, buy them along with the tube. In a pinch, you can use ¼-inch fuse clips from Radio-Shack (PN# 270-739). See FIGS. 14-2 and 14-3. The clips are attached to a bakelite base. Leave the clips attached to bakelite and cut it in half so you have a clip on each side. Then bend the clip a little to adjust it to the right size to fit the laser tube terminal. Solder your lead wires to the clip and place them on the laser tube terminals (see FIG. 14-4).

If you are powering a laser head (a laser tube enclosed in a housing), identify the polarity by the color of the leads coming out of the housing (see FIG. 14-5). Red denotes positive (anode), and black or sometimes white denotes negative (cathode).

Adjusting the power supply with a VOM

If you do not have a VOM, it is still possible to adjust the power supply. Read through this section anyway, because it gives pertinent information on adjusting, and then you can jump to the section on adjusting the power supply without a meter.

14-3 Modified Radio-Shack fuse clip.

14-4 Spring clip connection to laser tube.

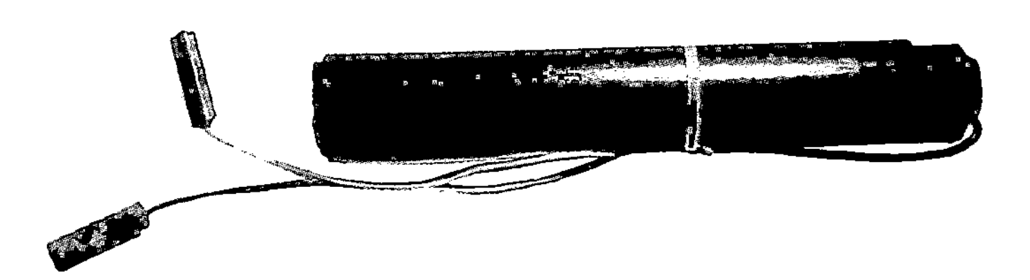

14-5 Laser head.

Connect a VOM as shown in FIG. 14-6. Set the meter to read milliamps. Turn on the power supply. Adjust capacitor C2 using a screwdriver until the reading on the VOM matches the laser tube's recommended power in milliamps. When adjusting the power supply, use a screwdriver with a nonconductive handle. If you touch the capacitor, it will throw the circuit timing off, cutting off or underpowering the laser tube and giving you bad VOM readings. If it's unavoidable that you touch the capacitor, either from holding it while adjusting or with the screwdriver, make small adjustments and release the capacitor completely to read the meter. Keep adjusting in this way until the VOM reads the proper current required by the tube. Once the power supply is adjusted, let it run for an hour or so as a break-in period. Afterward, recheck the VOM, and readjust the power supply if necessary. The power supply is now complete, so remove the VOM from the circuit—but don't forget to short the output of the circuit before removing the meter or you'll get a shock.

Some of the components run a little warm, but this is not a cause for concern. The transformer, rectifier, and ballast resistors might feel a little hot.

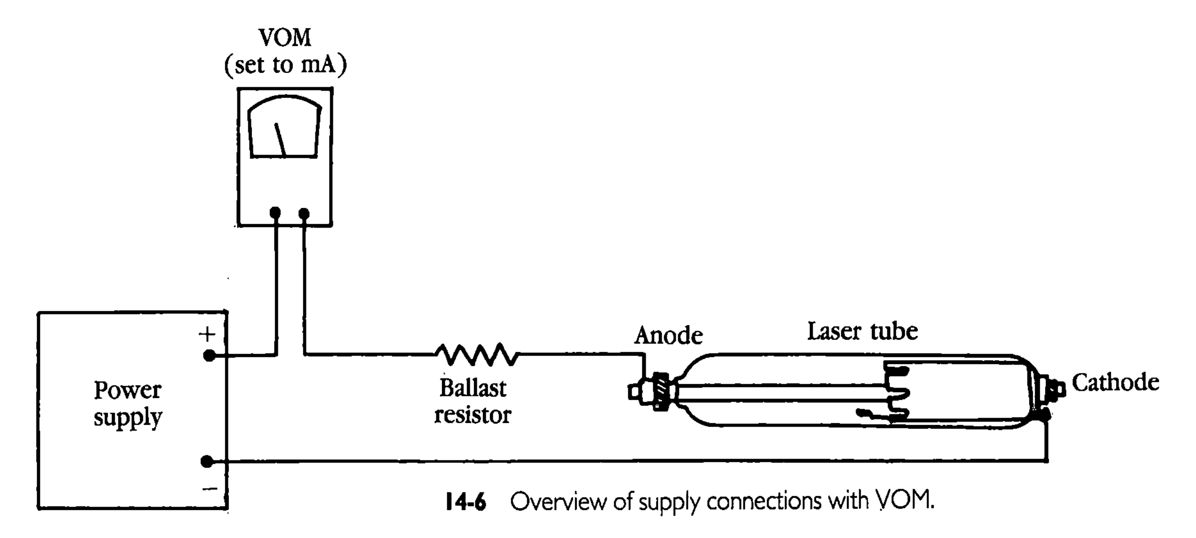

14-6 Overview of supply connections with VOM.

Adjusting the power supply without a meter

First let me throw a pitch for you to buy a meter. VOMs are inexpensive, less than $20 at Radio-Shack. You will find the VOM invaluable in all electrical work.

Adjusting the power supply without a meter is like flying by the seat of your pants. It's not exact, but it'll get you where you want to go. The adjusting capacitor C2 provides full power to the laser tube when its adjustment screw is turned all the way in. Before starting the power supply, make sure the screw is turned all the way in. Make all the connections to the laser tube, and don't forget the laser tube's ballast resistor. Turn on the power supply, and back off the adjustment screw on the capacitor (C2) until the laser begins to sputter. Reverse direction and turn the screw on C2 back in until the laser begins to function properly and then stop. Now, continue to turn the screw another ¼ turn in and stop. This adjustment will bring you in the ball park and allow you to operate the laser. Be aware that you might not be supplying the laser tube at its optimum power, which could shorten its life.

TROUBLESHOOTING

This section lists some of the common problems in shooting holograms and some possible solutions.

Too much power

If you cannot reduce the current to the proper level required for your laser tube, add a few ballast resistors into the circuit. Note this is not the laser tube ballast resistor. This is the 1-ohm, 10-watt resistor marked R4 in the power supply schematic. Add one or two resistors in series until you bring the current level down to where it needs to be.

Too little power

Insufficient current to the laser tube will cause it to sputter (see description below). To increase the current capacity of the circuit, remove resistor R4 from the circuit (see schematic). If this change doesn't supply sufficient current, use a larger capacity transformer for T1. A higher capacity transformer does not mean higher voltage; it means that it can supply more amps at the same voltage. By making this change, however, you will probably have too much power (see previous section).

Sputtering

A laser tube that sputters (turns on and off rapidly) probably isn't receiving sufficient current. Check all the power supply connections. Make sure the power supply isn't arcing. Double-check the ballast resistor, make sure you are using the proper value of ballast resistor. Check the distance of the ballast resistor to see if it is close enough to the anode of the laser. Excessively long leads to the laser tube could cut the power supplied.

Bad laser tubes

If your tube refuses to function properly, you could have a bad laser tube. Check it by operating the power supply with just the laser tube ballast resistor. If you have a stable current flow in the proper range as read on a VOM, call the company you purchased the tube from, explain the problem, and return the tube for replacement.

POWER SUPPLY ENCLOSURE

Use a nonconductive enclosure, one that is either made out of wood or plastic. This is especially true for the power supply described here. Because the outer casing of the 12-volt ignition coil is an electric ground, if it's enclosed in a metal enclosure and makes metal-to-metal contact with the enclosure, it could short out the circuit or create a shock hazard.

The on/off switch is a push-on/push-off type of switch. The "on" indicator light is a green neon lamp. Do not substitute any other color neon lamp. The green color was chosen to have a minimum impact on the film used to make holograms (see chapter 5). The switch and light are mounted on the top of the enclosure.

SURPLUS POWER SUPPLIES

The laser power supply has a required input voltage. Some are available that operate directly from your household electricity (117 Vac). If given the choice, this is the one to purchase.

Others require a 12-volt dc input of 12-volt ac input. Although this isn't a problem, both voltages require you to buy a 120 V/12 V step-down transformer. And in the case of 12-volt dc input, a rectifier and capacitor is also needed. The schematics for all three surplus power supplies are illustrated in FIG. 14-7.

The power supply typically looks like a black rectangular box with wires coming out of it. The input and output wires are clearly marked on the box. On one side of the box, you will find an adjustment screw that varies the output power of the power supply to adjust it to whatever type of laser tube you are using. After the unit is wired and connected to the laser tube, follow the

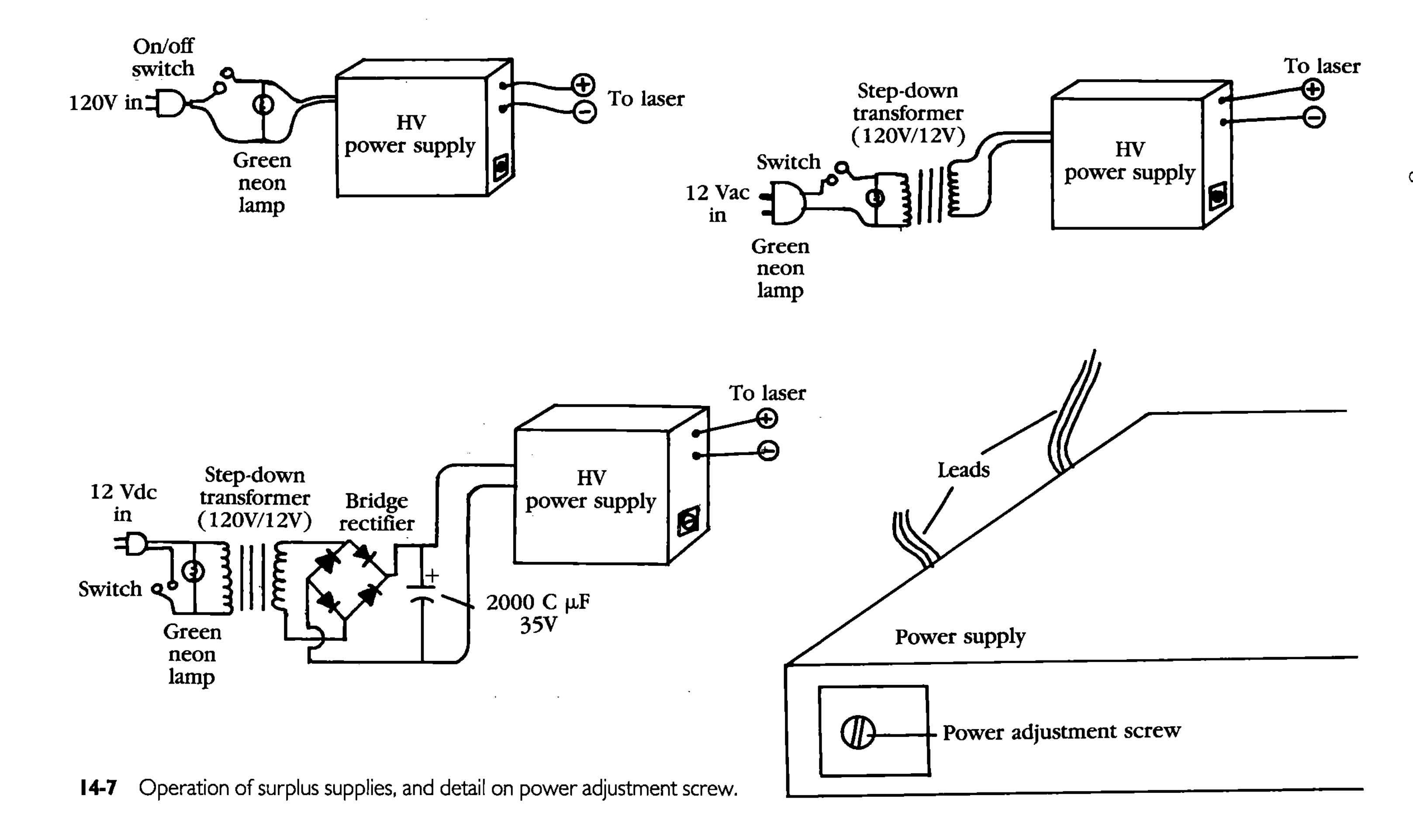

14-7 Operation of surplus supplies, and detail on power adjustment screw.

instructions for adjusting the power supply to the laser given for the fully constructed power supply.

You will notice that all power supplies use a green neon indicator light. Do not substitute any other color light because the green light will have the minimum impact on fogging the holograph film (see chapter 5).

All information and precautions given on the homemade power supply also pertain to these surplus power supplies. So if you have skipped over that information, go back and read it. The only section you might not need is that on construction.

LASER TUBE HOUSING

After you have run all the preliminary checks on the power supply, you should construct a suitable mounting and enclosure for the laser tube. Figure 14-8 illustrates one mounting system. I was fortunate enough to locate plastic clamps with the same diameter as my laser tube. I mounted these to a wood base and secured my laser tube in them.

14-8 Mounted laser tube.

You will probably have to construct a simple mount system. The easiest way to do this is illustrated in FIG. 14-9. Get a piece of lumber 4 × 3 × ¾-inches thick. Drill a hole in the center of the wood the same diameter as the tube or laser head you want to mount. Large-diameter cut-out drills are commonly available in hardware stores to drill doorknob openings in doors. If they don't have the exact diameter you need, take the closest diameter that is larger than the diameter you need. After you have drilled the hole, cut the wood in half so that you have two semicircles on each piece. Mount these pieces to a wood base as illustrated in FIG. 14-9. They are secured to the wood base by using wood screws that come up from the base into the wood pieces. To make the unit more secure, you can apply wood glue before assembly. Place the laser tube or head into the wood pieces as shown. On each side of the wood, place a wood screw as shown in the detail drawing. Using 22-gauge solid, insulated wire, wrap the wire around the wood screw and lash it across the laser tube. Secure it to the other side using a wood screw in the same manner. Do this with both mounting pieces of wood.

If you are using an enclosed laser head as in FIG. 14-5, you're just about finished. Make the connections to the power supply. Make sure all connections are well insulated and secure. You're ready to go.

If you are using a standard laser tube, you must build an enclosure. The reason for this is once the tube is powered, the glow of light from the tube will

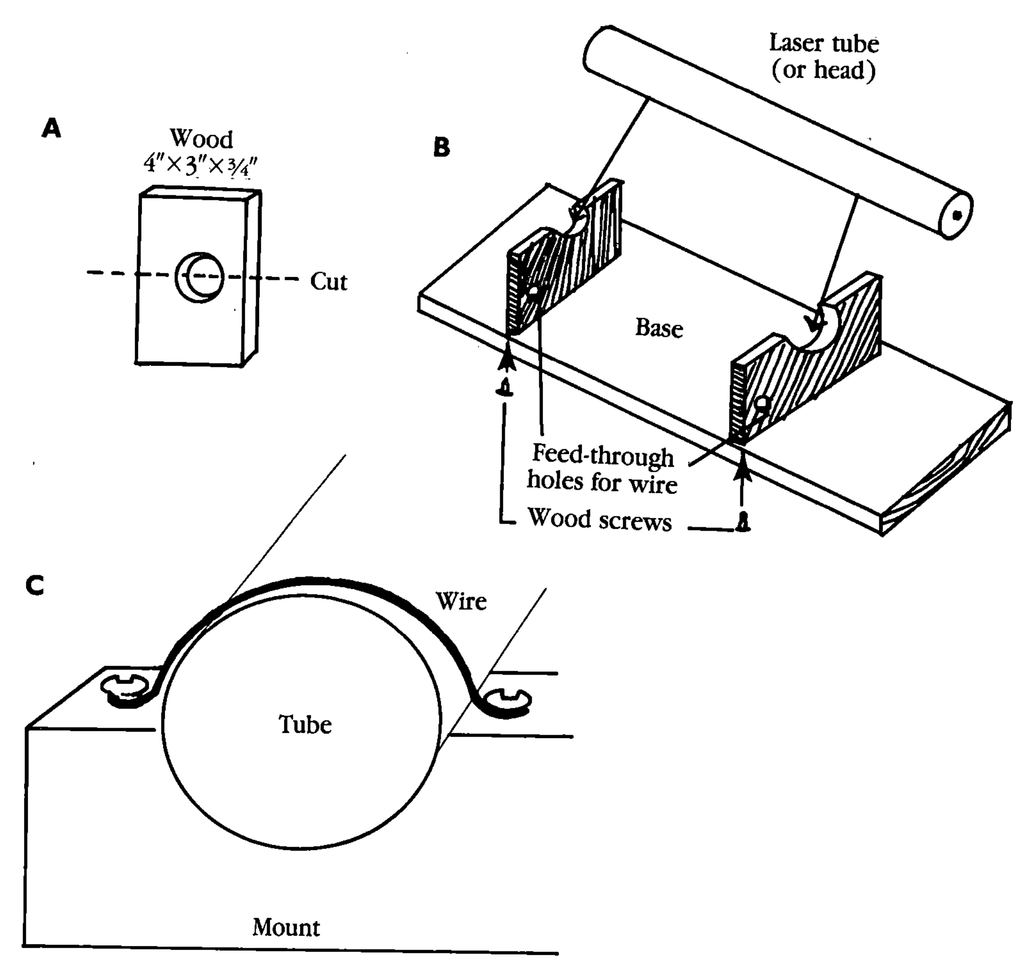

14-9 Construction of a simple laser tube (or laser head) mount. (A) Cut the wood to size and cut a hole the same diameter as the tube or head; then cut in half crosswise. (B) Position pieces on a wooden base to determine where to cut feed-through holes for wire; cut holes. Mount pieces to base with glue and wood screws. Place tube into holders. (C) Secure tube to holders with 22 gauge wire strapped over the tube and wrapped around and secured with two wood screws.

expose any film you're going to shoot, so the enclosure blocks all that light. Drill a hole in one end of the enclosure to allow the laser beam to pass through. Just about any enclosure can be used. Mount the tube assembly inside the enclosure, and on one end, place a small hole for the laser beam to pass through.

I used a rectangular aluminum pipe, but I did encounter a problem. Because of the close proximity of the tube to the walls of the enclosure, the tube electricity arced to the walls when I started it. I solved this problem by masking the inside of the tube near the cathode end with electrical tape.

Appendix A

Laser safety

For the HeNe lasers used in this book, you must be concerned with eye safety (although these lasers are totally skin safe). The American National Standards Institute (ANSI) has determined what power levels and types of laser light are considered safe for eye exposure.

RADIOMETRIC MEASUREMENTS

To determine the power levels of radiated light, first look at the inverse square law. Figure A-1 illustrates a light source that illuminates 1 square centimeter of space with 1 watt at a distance of 1 meter. If that 1 watt of power continues to propagate until it covers 4 square centimeters at the 2-meter mark, the power per square centimeter is ¼ or 0.25 watt per square centimeter. If the light radiation continues to propagate until it covers 9 square centimeters, then the power would be ⅑ or 0.11 watt per square centimeter.

With lasers, a lens is typically used to spread the beam. If you know the power and the diameter of the spread beam, you can calculate the beam intensity by dividing the beam power by the area it covers. The result is in power per square centimeter since that is how most power calculations are expressed.

AREA OF A CIRCLE

The spread laser beam typically illuminates a circular area, so first calculate the area of a circle.

$$A = \pi r^2$$

where A = area in square centimeters
π = pi = 3.14
r = radius

An unspread laser beam typically has a diameter of 0.5 to 1.5 millimeters. Assuming your laser has a beam diameter of 0.8 mm, the area covered is:

$$\begin{aligned} A &= \pi r^2 \\ &= 3.14 \times (0.4 \text{ mm})^2 \\ &= 3.14 \times 0.16 \\ &= 0.50 \text{ mm}^2 \end{aligned}$$

To convert millimeters to centimeters, divide by 10:

$$A = 0.05 \text{ cm}^2$$

ANSI SAFETY STANDARD

ANSI uses the nomenclature "MPE" to denote *m*aximum *p*ermissible *e*xposure. The MPE values are below known hazardous levels. The exposure is determined by time, *t*, in seconds and multiplied by the intensity of the laser light. For prolonged viewing, a maximum energy of 17 μW per square centimeter of laser light is considered safe.

Aversion response indicates the MPE of accidental exposure. The power level of the laser light is based on an aversion response time of one-fourth of a

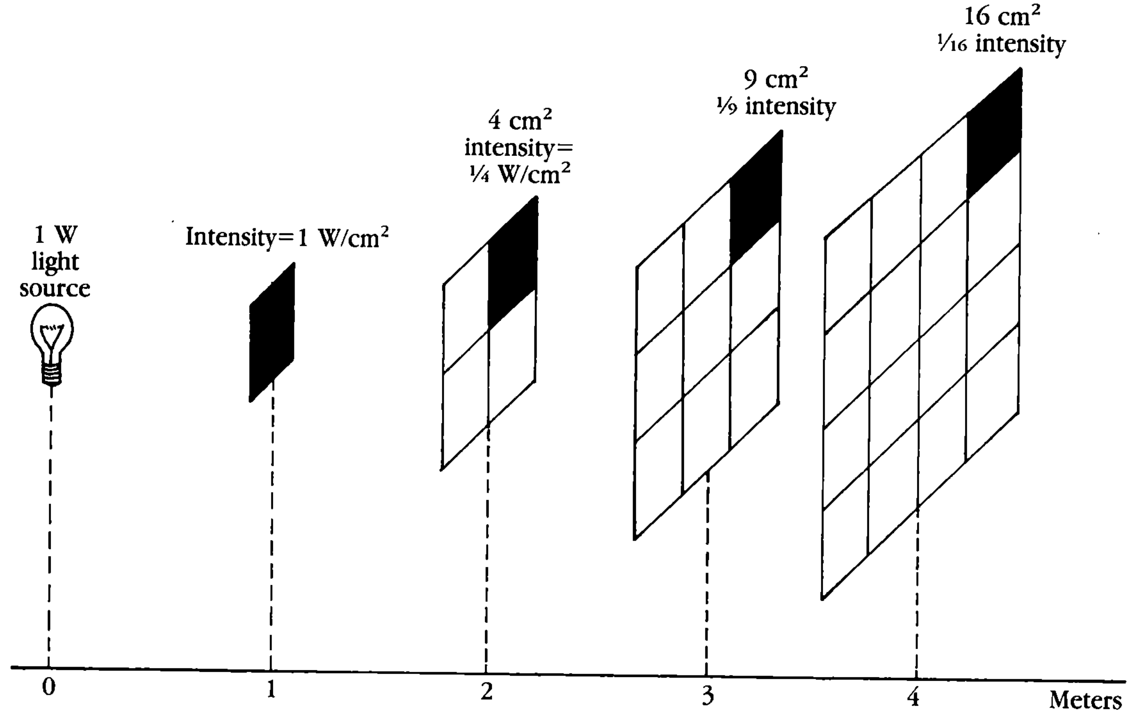

A-1 Inverse square law. As light spreads outward, the intensity of the light is inversely proportional to the square of the distance. For example, at a distance of 3 m, the intensity of light is 1/9 the intensity at 1 m.

second. This is the typical time required for you to avoid a laser beam from accidentally striking your eye by closing your eye and moving your head. The MPE for aversion response is 2.5 mW per square centimeter.

SPREAD-BEAM CALCULATIONS

The following examples use a 5 mW laser. Figure A-2 illustrates a laser beam spread with a negative lens. Remember that the energies are calculated by dividing the laser beam strength in mW by the area of the illuminated circle in cm^2.

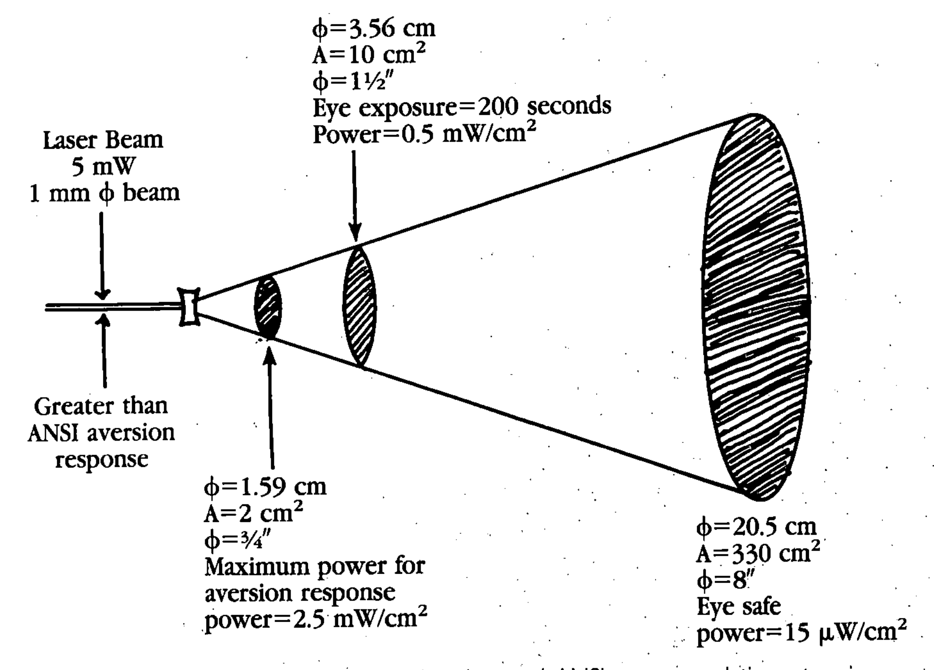

A-2 Spread laser beam, with power ratings and ANSI recommendations at various diameters.

A 5 mW unspread laser beam, with a diameter of 1 mm is well above the ANSI safety standard for aversion response. However, even a 0.5 mW beam with this diameter is above the standard, avoid looking directly into an unspread beam regardless of the laser power.

The situation changes dramatically as the beam is spread. The maximum power for aversion response is at a diameter of 1.59 cm. Continuing further to the diameter of 3.56 cm, the maximum exposure at this power level is 200 seconds. Finally, the eye safe level is at 20.5 cm (8 inches). If your laser is lower in power, the spread-beam diameters will be somewhat smaller.

To calculate the beam strength at any illumination point, first calculate the area covered by the laser in square centimeters (cm^2). Divide the laser power in milliwatts (mW) by the area to arrive at the power per square centimeter. Using the equation for the area of a circle, recalculate the examples given in the drawing.

OTHER USES

Aside from keeping your work area eye-safe, these calculations do have other uses. First, if you build the laser power meter, you can use the calculated beam energies to calibrate the meter. Secondly, all holographic films are rated at different exposure energies. You can use the calculations to estimate the power at the film plane and adjust your exposure accordingly. Don't get discouraged if the exposure values are off; holographic film is notorious for having inconsistent exposure energies that vary from batch to batch.

Appendix B

Miscellaneous projects

This appendix includes plans for the audible timer, electronic shutter, laser power meter and photometer, spatial filter, and hints on cutting glass.

AUDIBLE ELECTRONIC TIMER

Since I do most of my development work alone, a problem I had when developing holograms was to keep track of the time the film plate was in the developer solution. Rocking the tray to keep fresh solution in contact with the plate keeps both hands occupied, and wearing gloves makes it difficult to look at a wrist watch. Standard darkroom clocks with large luminescent hands work fine but are expensive (I couldn't find any for under $80). So I decided to build my own timer.

The timer I designed is audible; it beeps every 15 seconds. The schematic for the timer is simple and fits into a small box (see FIG. B-1; parts list is in TABLE B-1). The LED used in the circuit is a subminiature green LED. Do not substitute any other LED for this one. The subminiature green LED was chosen for two reasons. First, because of its small size, it takes a minimal amount of current to light, which prolongs the battery life. Second, if you use a standard-size LED, there isn't sufficient circuit current to light it. It should be green so it doesn't fog the holographic plate.

Using the timer

When you first turn on the timer, it takes longer than 15 seconds for the first beep to occur because the timing capacitor is charging from 0 volts and therefore takes a little longer to breach the threshold of the timer. After the first beep, all other beeps will be spaced equally at approximately 15 seconds. The time is *approximately* 15 seconds because of the tolerances of the resistor and capacitor values (± 10 percent). Regardless, the time space is close enough to 15 seconds to be used reliably. Since the plate stays in the developer for about

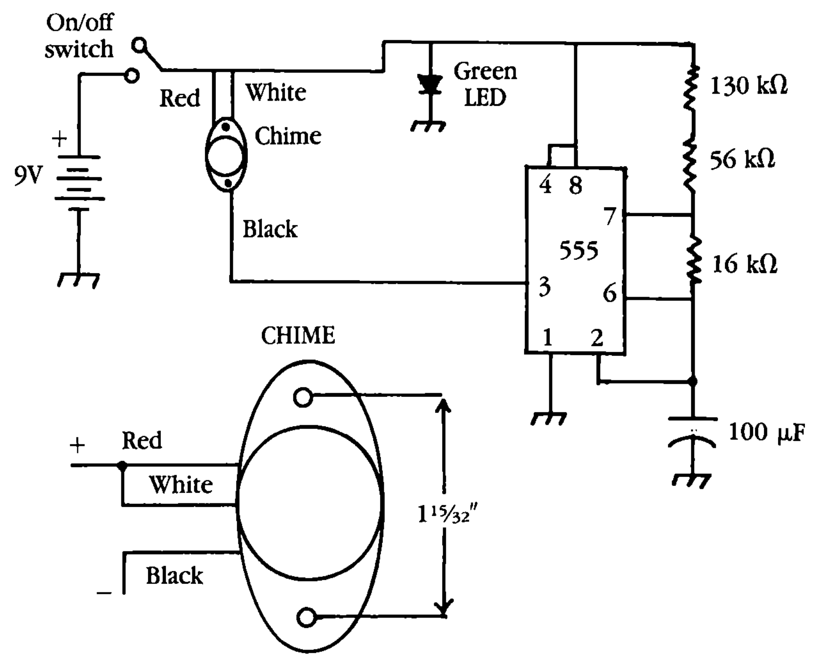

B-1 Schematic for audible timer. Timer chimes every 15 seconds.

Table B-1 Parts List for Audible Timer

Component	*Radio-Shack part number*
Electronic chime	273-071
555 timer	276-1723
Subminiature green LED	276-037A
100 µF capacitor	
Battery	
Case	
Resistors: 130 kΩ, 56 kΩ, and 16 kΩ	
9V battery	
Toggle switch	

NOTE: Do not substitute 555 IC timer with CMOS version.

two minutes, this is equal to eight beeps. If you have neutral density filters attached to your safelight, start checking the density of the hologram after four beeps (1 minute).

Construction is straightforward. Point-to-point wiring is fine. You could reduce the cost of the unit substantially by replacing the electronic chime with an inexpensive buzzer (also sold by Radio-Shack). Personally, I didn't like the sound of the buzzer.

ELECTRONIC SHUTTER

The electronic shutter is much more convenient than using a shutter card. With a shutter card, you must lift the card, wait to expose your plate and return the card to its position. As stated previously, just the process of lifting the card introduces vibration on the table. This can be eliminated by using the electronic shutter. When I tested the shutter on the table with an interferometer, the shutter had zero impact on the fringe movement (see chapter 2). An added advantage is that the shutter can be controlled by a simple timer circuit. The timer circuit is much more reliable than counting off seconds during the exposure. It makes it easier to fine tune the exposure to give you the best image. Building the shutter and the timing circuit might appear a little intimidating to the novice, so a simple battery control is shown if you would like to start with that. When you feel you want better control over the exposure time, you can build the timer circuit.

For anyone trying to shoot holograms on a wooden floor, the electronic shutter becomes imperative. Any body movement or shifting as you move the shutter card in and out of place is sure to introduce vibrations through flexing of the wooden floor. Verify this by testing the area with the interferometer. If you use the electronic shutter, you can set up the table, move 5 feet away from it, wait the appropriate relaxation time, then expose the plate using the electronic shutter with either the small hand controller or timer. Since you are away from the table and simply have to push a button to make the exposure, the introduction of vibration through the floor is minimized.

Construction

The heart of the shutter is a voltmeter, available at Radio-Shack (part number 270-1754). On the back of the meter are two brass, threaded rods with nuts for mounting the meter to a panel. Remove the nuts and cut off the threaded rods so the back of the case is flush. Next, drill a 3⁄16-inch-diameter hole through the back of the voltmeter and through the transparent front piece of the voltmeter, as shown in FIG. B-2. Even though the front piece is made of transparent plastic, a laser beam passing through it will pick up a lot of added noise, so drill through it. Locate the hole in line with the indicator needle position of the meter when at rest. The hole should be at the midpoint of the indicator needle. Glue a small square or round piece of black opaque plastic to the indicator needle that will block the hole drilled in the case. Make sure the material is truly opaque and completely blocks the laser beam before gluing it to the meter needle (it would be really frustrating to go through all this trouble and find that the shutter doesn't block the laser light sufficiently). After gluing the piece on, clip off any excess needle that protrudes above the opaque material to lighten the weight of the needle. Do not glue a very large or heavy piece of material on the needle because such meters are not designed to move much weight.

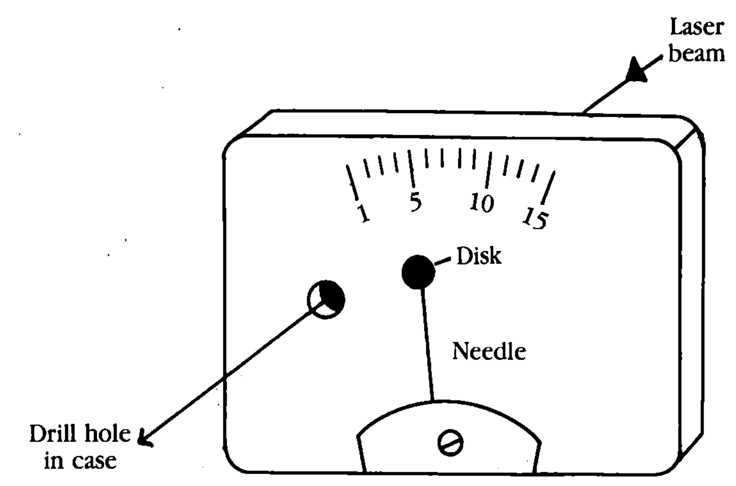

B-2 Illustration of voltmeter face and hole placement. Attach a small, opaque piece of plastic to the needle to block the laser unless the meter is activated.

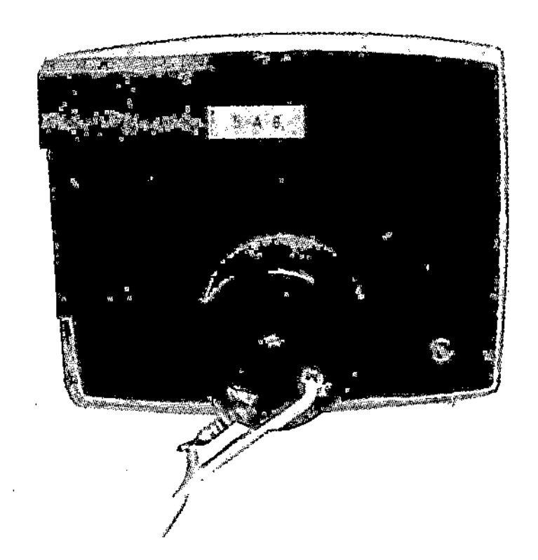

B-3 Back of voltmeter showing magnet placement.

The meter is supplied with a 15 kΩ resistor. Solder this resistor to the positive terminal of the meter per the instructions. Solder about 5 feet of wire to the negative terminal and to the free end of the 15Ω resistor.

Glue two magnets on the back of the meter as shown in FIG. B-3 to enable you to position the meter on the steel plate as with the other isolation table components (see FIG. B-4). The magnets are available from Images Co.

Simple shutter control

The simple control is a 6-volt battery and a push button installed in a case. The push button should be a normally open momentary contact switch. The leads from the meter connect the the unit as shown in FIG. B-5.

To use the shutter, align it with the laser beam so that the beam goes through the back of the meter but is blocked by the opaque plastic on the

B-4 Shutter on optical stand.

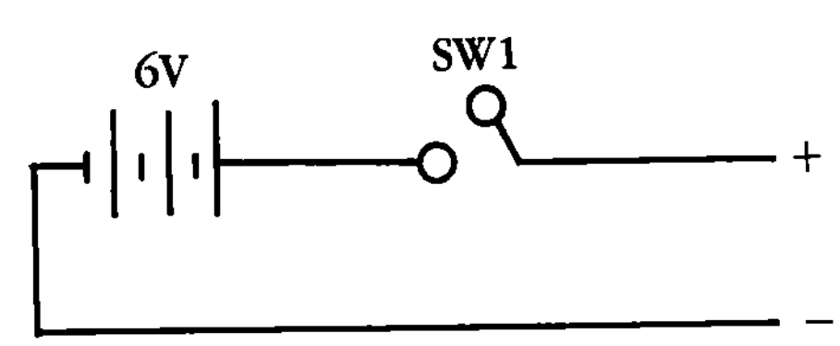

B-5 Schematic of simple shutter control. SWI is a normally open momentary contact switch. For 6V, use four AA batteries in a holder. Mount components in a small case and run the leads to the shutter.

needle. When you activate the meter, the needle swings up, unblocking the laser beam to make the exposure.

Electronic shutter timer

The shutter timer is another simple circuit (see FIG. B-6; parts list is in TABLE B-2). The case appears larger than what you would expect from such a small circuit. This space is merely to hold the push button switches and timing potentiometer. The timer is adjustable from 0.25 seconds to 20 seconds in two ranges. The

Table B-2 Parts List for Electronic Shutter Timer

Quant.	*Component*	*Radio-Shack part number*
3	Miniature push buttons (N.O.)	275–1547
2	10 μF capacitors	272–999
1	555 timer	276-1723
2	SPST miniature switch	275–624
2	Green subminiature LED	276–037a
1	Case	270–284
1	Push-in terminals	274–621
1	2.2 kΩ resistor	
1	22 kΩ resistor	
1	10 kΩ resistor	
1	5.6 kΩ resistor	
1	9V battery and holder	

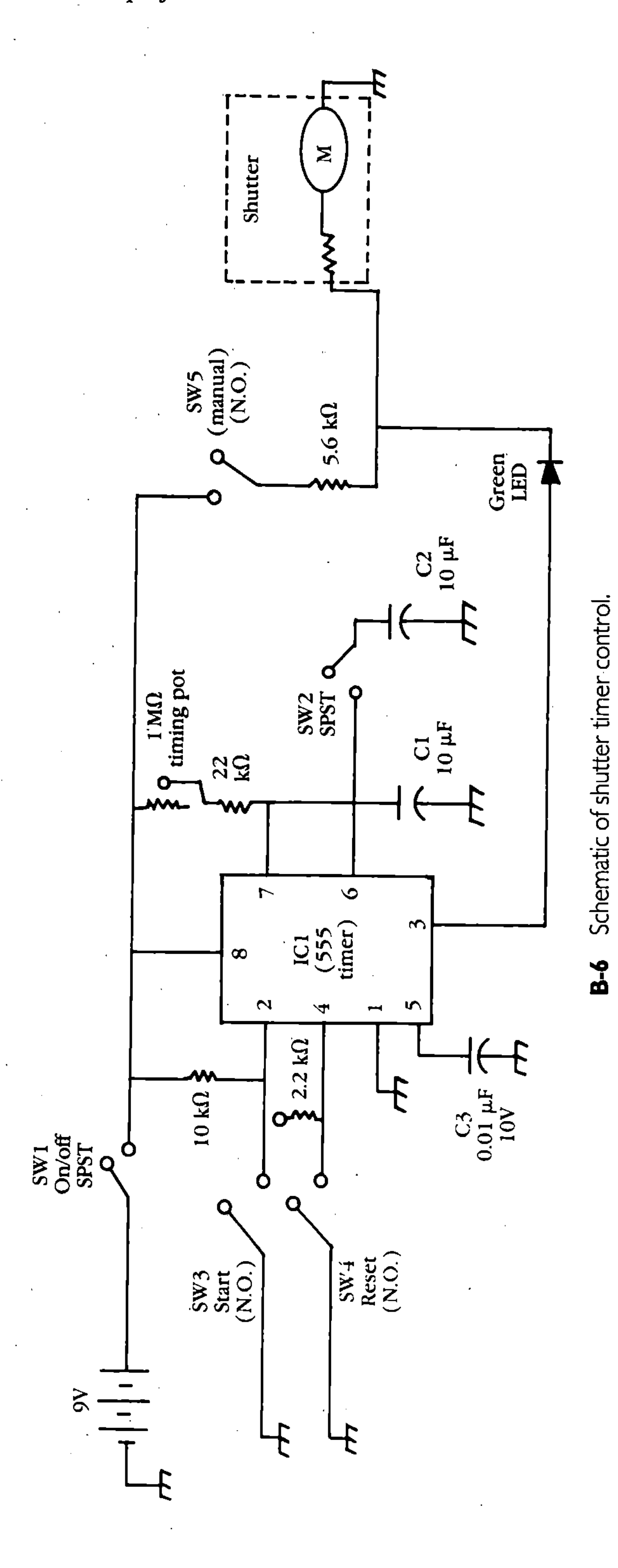

B-6 Schematic of shutter timer control.

first range is from 0.25 seconds to 10 seconds; the second range is implemented by the ×2 switch, which doubles the time indicated in the first range.

The push buttons used are normally open, momentary contact. They perform the following functions.

- START starts the timing cycle.
- RESET stops the timing cycle currently in progress.
- MANUAL activates the shutter manually. It can be used for aligning the meter or exposures longer than 20 seconds.

Construction is simple and straightforward. Attach push-in terminals to the front panel to connect the wire leads from the shutter to the timer (see FIG. B-7). This type of terminal is usually seen on the backs of speakers. The two LEDs are subminiature green LEDs and are used for the same reasons outlined for the darkroom timer (earlier in this appendix). One LED is an ON indicator, and the other lights when the shutter is activated; you can use the latter to calibrate the timer.

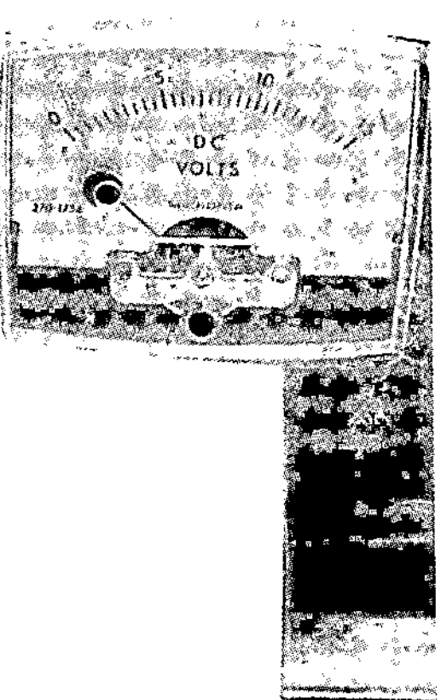

B-7 Overview of electronic timer and shutter.

Calibration

Once the unit is complete, calibrate the first range. Use either a stopwatch or the second hand of a watch. Starting with the longest delay, measure the time and mark it on the front panel at the current knob position. Continue to calibrate at times about 1 second apart. The ×2 range will be double the values you calibrated for the first range.

The timer, in combination with the shutter, is a useful tool. It allows you to easily reproduce your best exposure times. (As stated before, it becomes helpful if you keep thorough records on your holographic setups and results.)

LASER POWER METER AND PHOTOMETER

The laser power meter serves two purposes. It can be used to measure the output of your laser in milliwatts, or it can measure the light ratios in split-beam setups. See FIG. B-8. The parts list is in TABLE B-3.

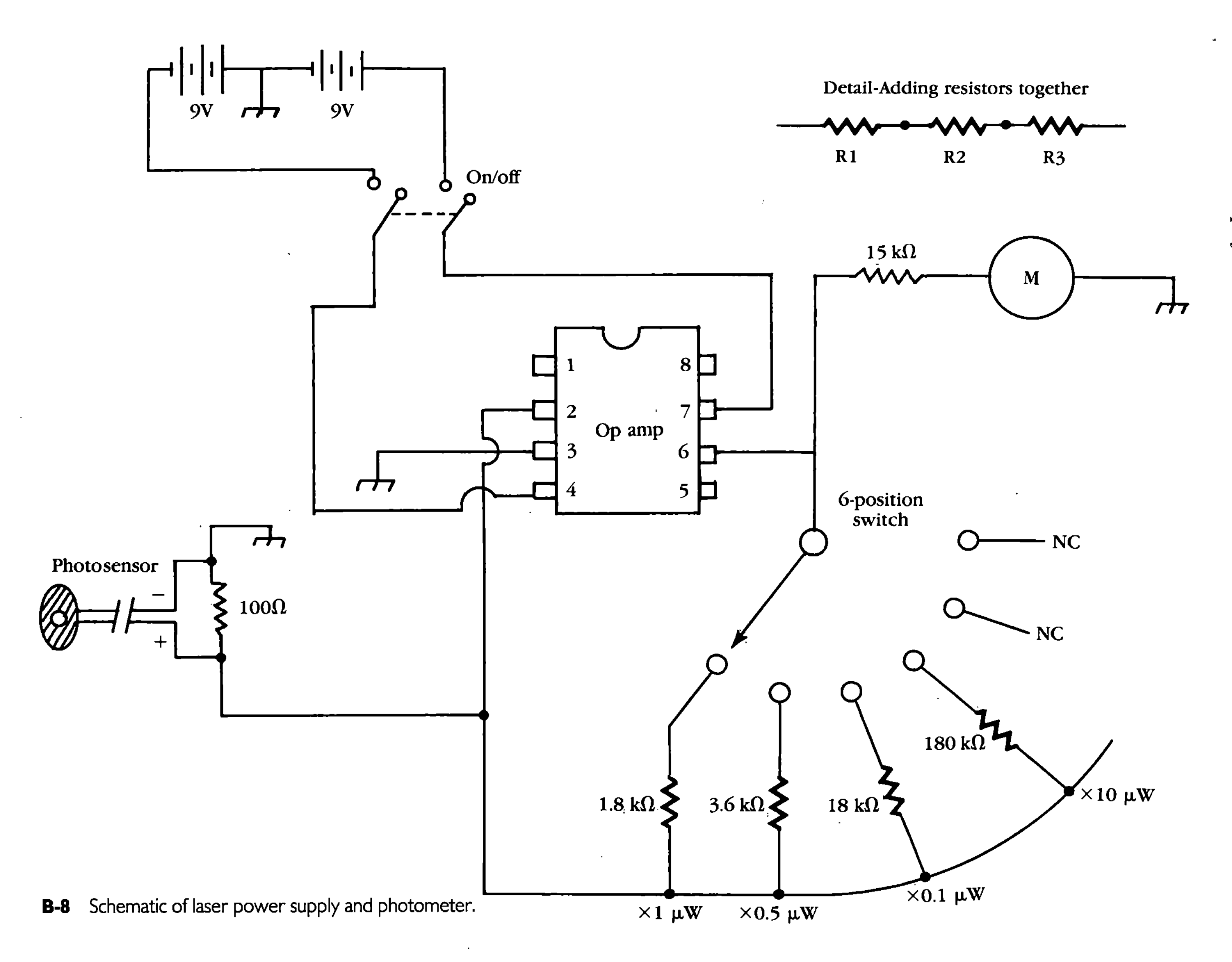

B-8 Schematic of laser power supply and photometer.

Table B-3 Parts List for Laser Power Meter

Component	*Radio Shack part number*
Meter	270-1754
Case	270-627
Switch	275-636A
Photocell	276-124
Six-position switch	275-1386
IC1 741 opamp	276-007
Resistors: 100Ω, 15 kΩ, and those for the rotary switch (see text)	
Two 9V batteries	

The circuit uses a dual-polarity power supply that is made from two 9-volt batteries. Note the on/off switch. This is a double pole (DP) switch, which simply means it's two switches in one. The switch has four terminals, two for each half of the switch. A double-pole double-throw (DPDT) switch is more common and has six terminals. Each switch has a double-pole position. If you use one here, you still only need to use four of the terminals on the switch.

Separately, the circuit also uses a six-position rotary switch. If you have never used a rotary switch before, it can appear confusing. The switch has six positions but contains twelve switches; two switches are activated simultaneously for each rotary position.

In the schematic, each position on the rotary switch is connected to a resistor. These resistors control the gain of the op amp, which in turn provides a wide range of measurable light power ranges. The required resistor values are not commonly available, so you must combine resistors to add up to the correct values. Use the following suggested list of common resistors that when added together result in the values required by the circuit:

Required	*Add*
1800 Ω	1 kΩ + 400Ω + 400Ω
3600 Ω	3.3 kΩ + 330Ω
18 kΩ	16 kΩ + 2.2 kΩ
180 kΩ	130 kΩ + 47 kΩ

Solder the resistors in series for each switch position as shown in the detail of FIG. B-8.

Construction hints

To minimize wiring, solder the resistors directly to the switch terminals. Solder the opposite end of all the resistors together and run one lead wire to pin number 2. This hint will save you a lot of work and should help prevent a potential rats nest of wires.

When you purchase the photocell, it comes in its own plastic case. Use the small plastic case as a photocell holder. Solder wires to the photocell per the instructions on back of the packaging. Construct the photocell holder as in FIG. B-9. Check the opaque material you use to be sure it is truly opaque.

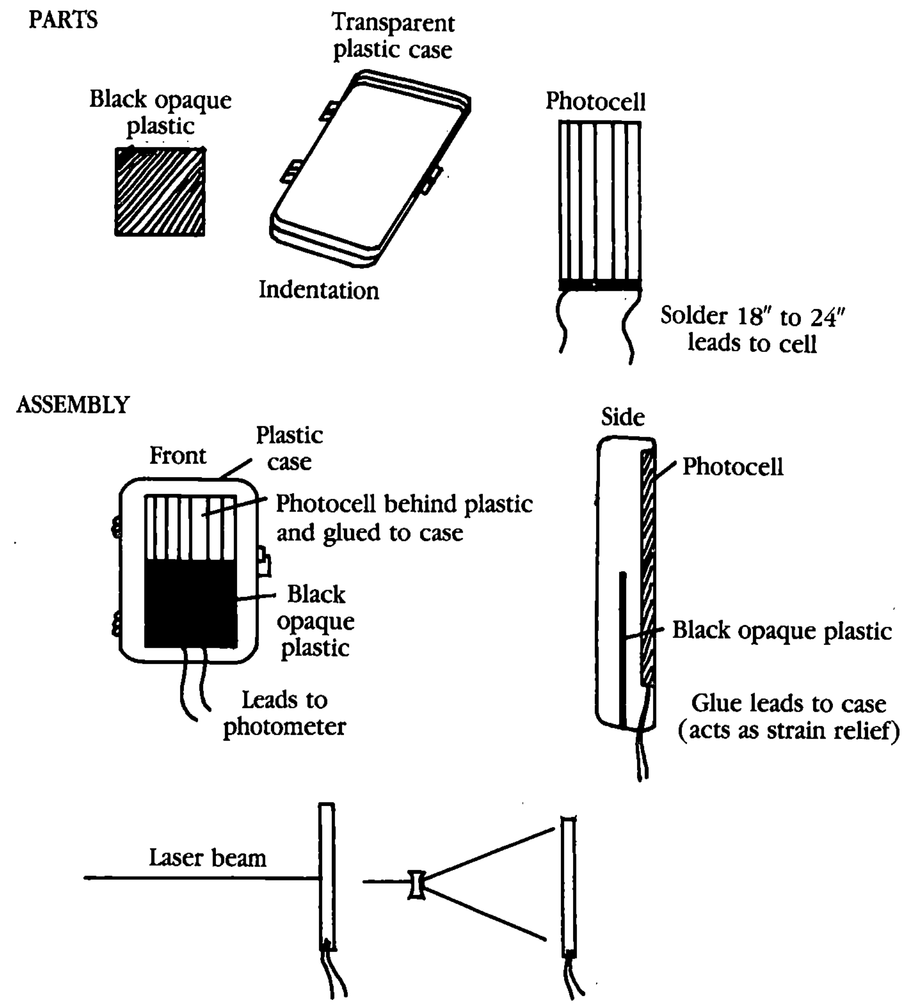

B-9 Construction of photosensor. Cut black opaque plastic to fit in case and cover ½ to ⅔ area of photocell. File an identation into bottom of case to allow leads to pass through. Solder leads to photocell. To assemble, glue the photocell to the plastic case behind the plastic, and glue the leads to the case. To measure light, hold the photosensor perpendicular to the light.

Using the meter

The meter is subdivided into scales. The first scale is labeled ×1 mW. Use this scale to measure the power output from your laser. Position the photocell in front of the laser beam so that the unspread beam falls perpendicularly onto the exposed photocell material. Multiply the reading in volts to mW. For example, if you read 3 volts, that would be equal to 3 mW. All other scales work similarly. You'll probably find the ×10 μW range useful for checking the beam ratios in split-beam setups. The range denotes power per square centimeter (cm^2). The procedure is the same for checking light power in split-beam setups with spread beams. Although the beam is spread, try to hold the photocell perpendicularly to it to measure.

There is one peculiar characteristic of the meter I would like to note. Say you turn on the meter with it set on a high range, like ×10 μW, and no light or insufficient light falls on the photosensor. The indicator needle swings backwards. This can be corrected by reducing the range position or putting the photosensor into the light.

The voltage range the circuit can supply is up to 8 or 9 volts; however, the voltmeter specified in Table B-3 is rated and scaled at 15 volts. The circuit will therefore never give a full-scale reading on the voltmeter. Ideally, we should use a 9-volt meter, but one is not commonly available, so I opted to a meter that is available from Radio-Shack. When using the meter, keep in mind that a reading of 8 volts could mean that you are over the maximum scale reading, so you should lower the range to the next position to get an accurate reading. You could mark the upper range between 8 and 9 volts with a red marker on the meter face as a constant reminder of this fact.

Accuracy

I was quite surprised by the accuracy of the meter in the lower scales when I compared it against a standard laser photometer. This does not mean that your meter will be as accurate as mine due to varying resistor tolerances. Consider the readings derived from the meter as estimates.

When you buy or build a laser, confirm its power rating with the meter, which should also give you an idea on the accuracy of it. New lasers typically put out more than their rated power to ensure the laser will supply its rated power output throughout its lifetime. For example, a new 4 mW laser often puts out 4.5 mW.

To ascertain the ratio accuracy of the meter, set up the laser and take a power reading from the beam. For the sake of illustration, say you measured 4.2 mW. Now place a 1:1 beam splitter in the beam path. Take readings from each split beam. Regardless of what the individual beam readings are, if they both add up to the original reading, the meter is fairly accurate. If one beam measures 2.6 mW and the other 1.6 mW, these add to 4.2 mW, which is quite accurate. You might discover that your 1:1 beam splitter isn't exactly 1:1. Usually, the transmitted beam is more powerful. You can also perform the same test with a spread beam to check the higher scales. At the higher scale, the accuracy decreases, but it is still useful for checking beam ratios.

To adjust the scale, increase or decrease the exposed area of the photocell using the opaque plastic in the photosensor (you need a commercial meter to calibrate).

Increasing sensitivity

If you're working on a split-beam setup, there might be insufficient light to measure using the photometer with the 10 μW scale. To increase the sensitivity, remove the opaque plastic from the photocell holder. Use the entire photocell to take a reading (this is only accurate for determining beam ratio strengths).

SPATIAL FILTER

The usefulness of this spatial filter is limited because of the restriction of using a large pinhole and a corresponding optical element. Despite the restriction, the spatial filter performs admirably well for the small film plates (2.5 square inch). It can be used for all single-beam setups and provides a very clean beam. The filter can also be used on a larger table to provide a clean reference beam in split-beam setups.

Figure B-10 illustrates the operating principle of the spatial filter. A lens focuses a laser beam to pass through a pinhole. Any light scattered by dirt or imperfections in the lens is effectively blocked.

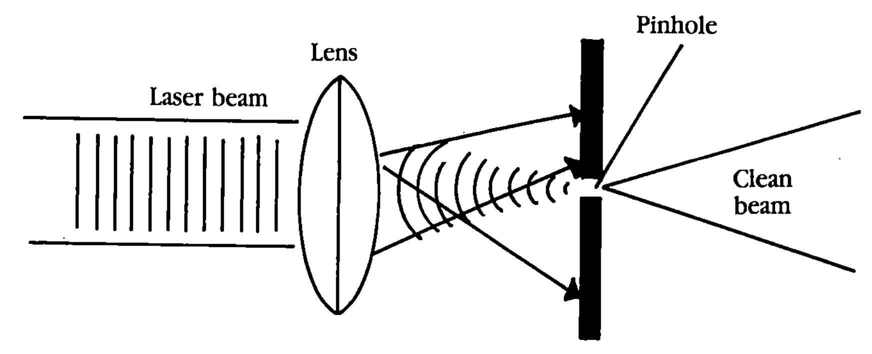

B-10 Operation of spatial filter. The arrows represent scattered light caused by dirt or imperfections in the lens.

To construct a spatial filter, gather the materials, as listed in TABLE B-4. Drill a ⅛-inch center hole in the aluminum plate, and a ⅜-inch hole in the steel plate. Cut the angle in half. As shown in FIG. B-11, rough up the mating surfaces of the plates and angles with sandpaper so the epoxy can get a good grip, and glue the plates to the angles. Glue the lens in the small hole in the aluminum plate, the same way as for the magnetic mounts.

Pinholes

You can make or purchase the pinhole. Pinholes are usually expensive. Edmund Scientific sells a kit of unusually sized pinholes for a fraction of their original cost. The kit contains three apertures (pinholes) in the following ranges: 5 to 12 μm, 20 to 35 μm, and 40 to 75 μm. (The Edmund part number is J36-379; cost is $36.95). The two larger pinholes can be used with the spatial

Table B-4 Parts List for Spatial Filter

DCX lens, 12 mm diameter × 9 mm focal length (Edmund Scientific)
2″ square × 1/16″ thick aluminum plate
2″ square × 1/16″ thick steel plate
4″ length angle, 1/2″ × 1/2″ × 1/16″ thick (steel or aluminum)
Pinhole (purchased or made)
Clay or Fun-Tak
Rectangular magnet

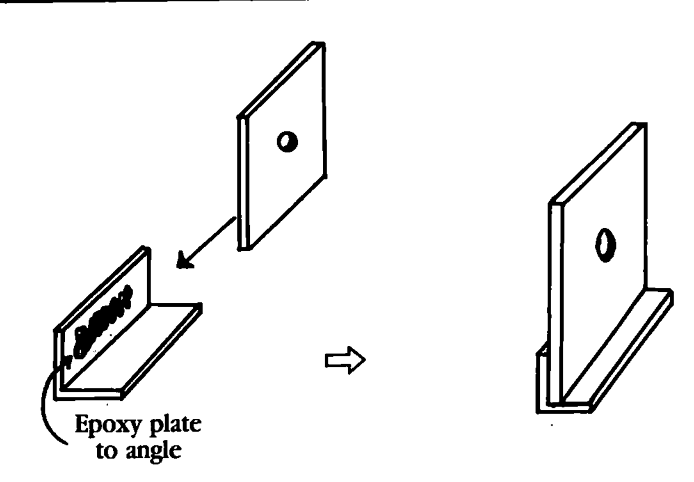

B-11 Construction of simple spatial filter.

filter outlined here. The smaller pinhole would take an eternity to line up properly.

Making a pinhole If you want to try your hand at making a pinhole, here's how to do it. Get some aluminum foil, the thinnest foil available, a small piece of glass (a microscope slide works well), and the smallest diameter sewing needle you can find. Cut ten 1-inch squares of aluminum foil. Stack them onto the piece of glass. Using the sewing needle, gently press down on the center of the stack with the needle so it goes through only about half the stack. Remove the needle from the foil. Start holding up the pieces of aluminum foil one by one up to a light. Looking at the light through the pinholes in the foil, select the foil with the smallest diameter (the piece on top of the one where no light is visible through it). You might have to try a few times to get a really small pinhole. When you have your foil pinhole, glue it to the rectangular magnet so that the pinhole lies centered with the hole in the magnet.

Aligning the spatial filter

You might want to start by trying to align the spatial filter on a table or bench rather than the isolation table so you can get the hang of it in a more comfortable position rather than being stooped over the isolation table. This can also help increase your patience, something you're going to need the first time you align this filter.

Figure B-12 illustrates the following procedure. With the laser on and pointed in the direction you want it to go, place a ½-inch-thick strip of clay on the table in front of the laser. Place the lens assembly in front of the laser and on top of the clay. Press the lens assembly into the clay and align the lens with the laser so that the spread beam illuminates the area you want. It's a good idea to place a white card or paper at this position to act as a screen. From this point on, do not touch the lens assembly any more. Take the pinhole that is attached to the magnet and place it in front of the steel plate so that the hole in the magnet is aligned with the hole in the plate. The foil pinhole should be on the opposite side of the magnet. Position another strip of clay about ½ inch in front of the lens assembly. Place the pinhole assembly onto this piece of clay. The aluminum foil pinhole should be facing the lens.

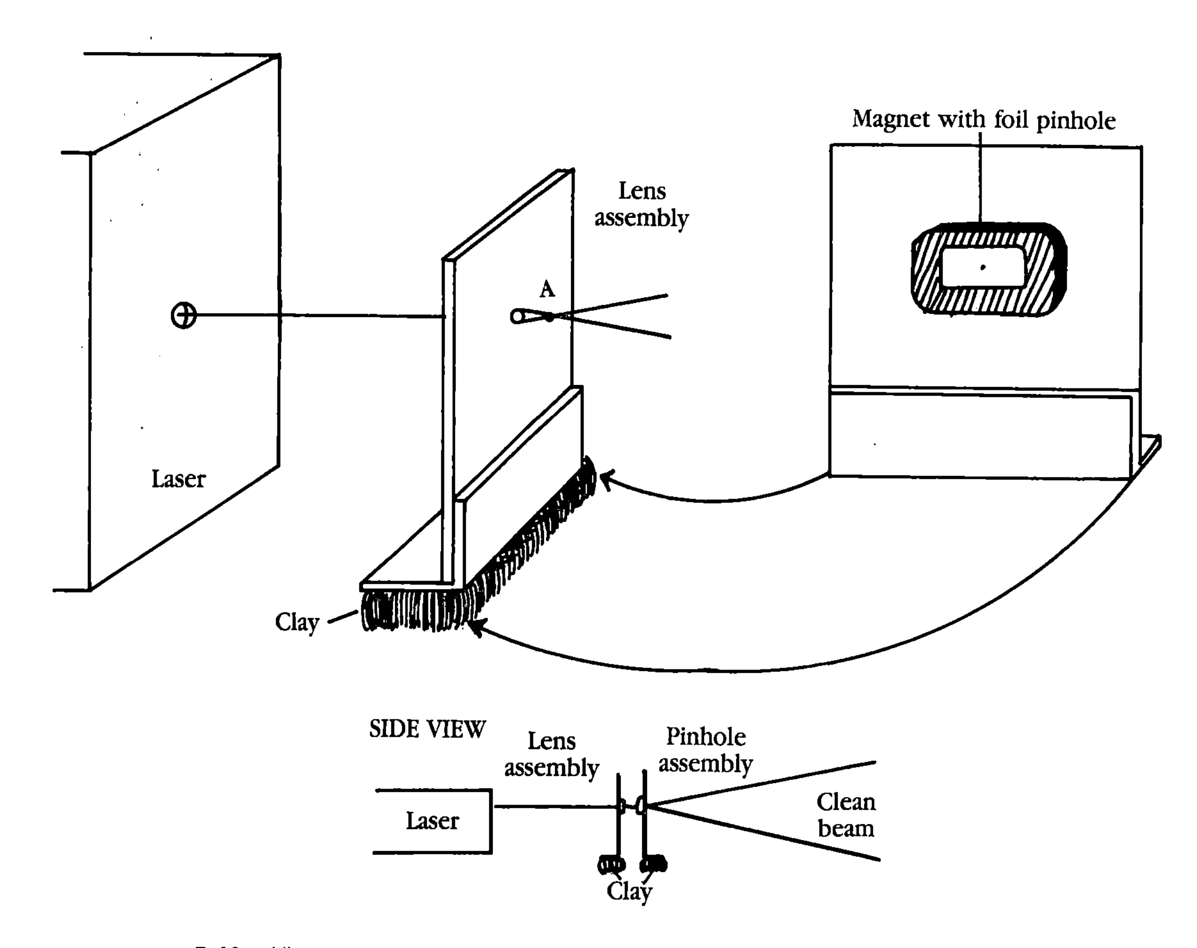

B-12 Alignment of spatial filter. Point A shows the location of the focal length of the lens; align the pinhole with it. For example, for a 9mm lens, the distance is about ⅜ inch.

Now comes the fun part. Align the pinhole with the focal point of the lens, which is 9 mm (about ⅜ of an inch). Positioning is easier if you dim the room lights. This makes it easier to see the light on the screen when you're getting close to the right position. Now move the pinhole around using the bottom angle piece. Keep your eye on the screen. Whenever you pass the pinhole position you'll see some light. When the pinhole is centered within the beam but is too close or too far away, a set of concentric rings appears on the screen (see FIGS. B-13 and B-14). When you reach the correct distance, an evenly illuminated disk of light will appear (FIG. B-15). This light should illuminate the screen almost as brightly as if the pinhole wasn't there. The first time I aligned the filter it took me fifteen minutes to get it right. With a little practice, you should be able to align the filter in less time. Although the photo doesn't do justice to how well the spatial filter cleans up the beam, FIG. B-16 is a photo of the lens operating alone; compare this to FIG. B-15, which shows the clean (filtered) beam.

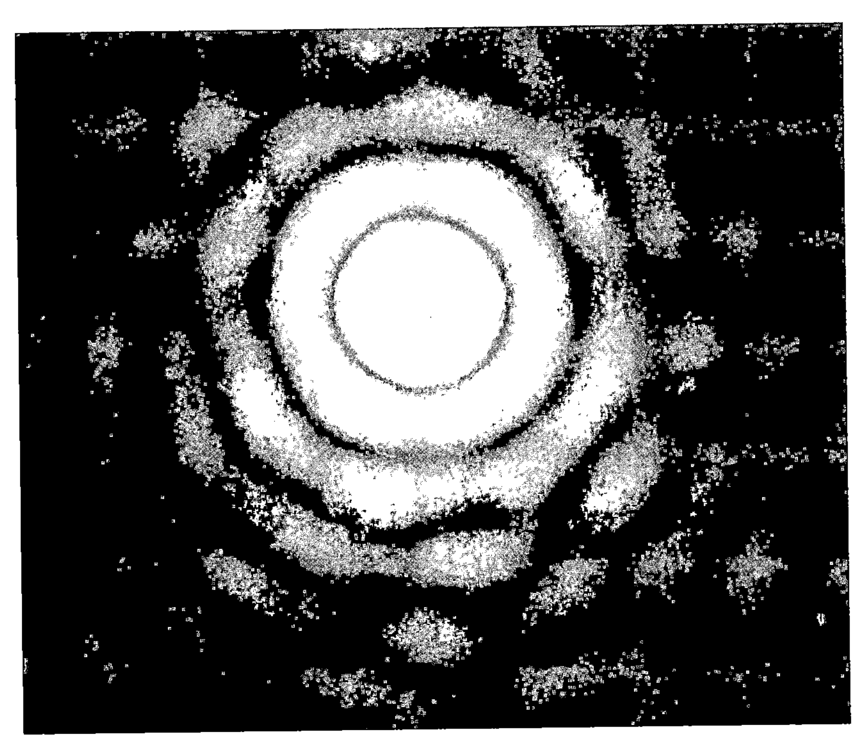

B-13 Airy disc created by spatial filter during alignment.

Matching the lens and the pinhole

Typically, a 25 μm (0.001-inch) diameter pinhole is about the best you can hope for when making a homemade pinhole. The DCX lens works with a 25 μm pinhole. A couple of simple formulas can help evaluate what size pinhole should go with any particular lens.

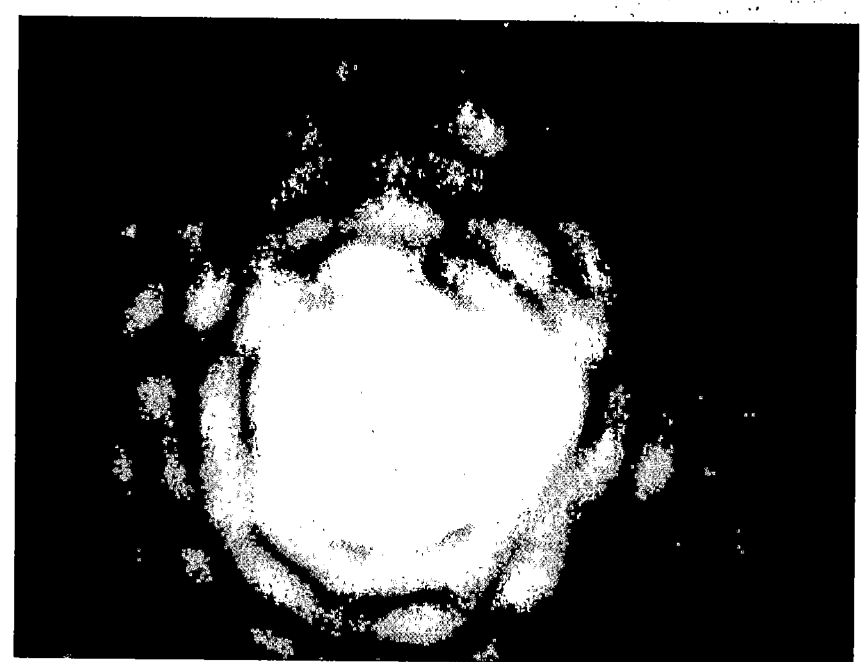

B-14 Progressing in alignment, the Airy disc begins to expand.

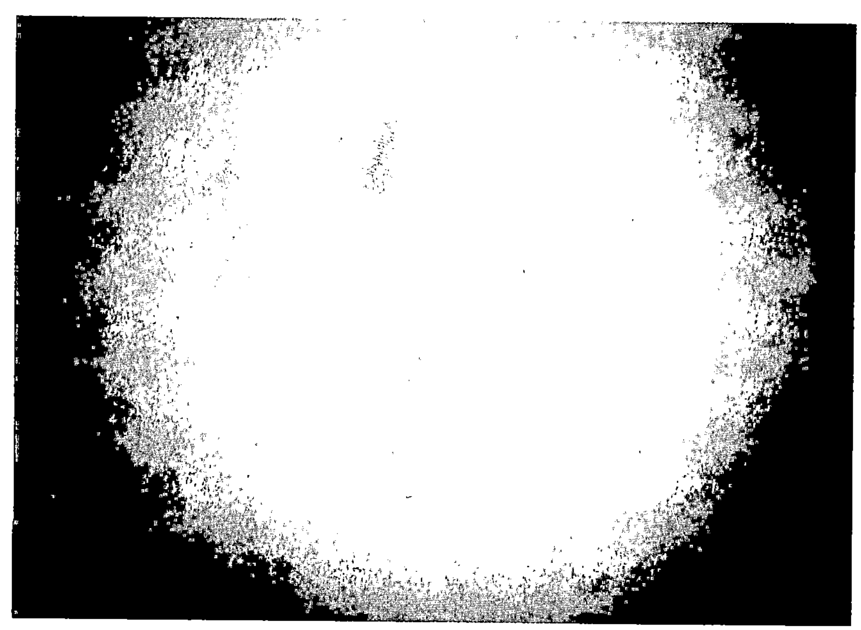

B-15 Beam spread from perfectly aligned spatial filter.

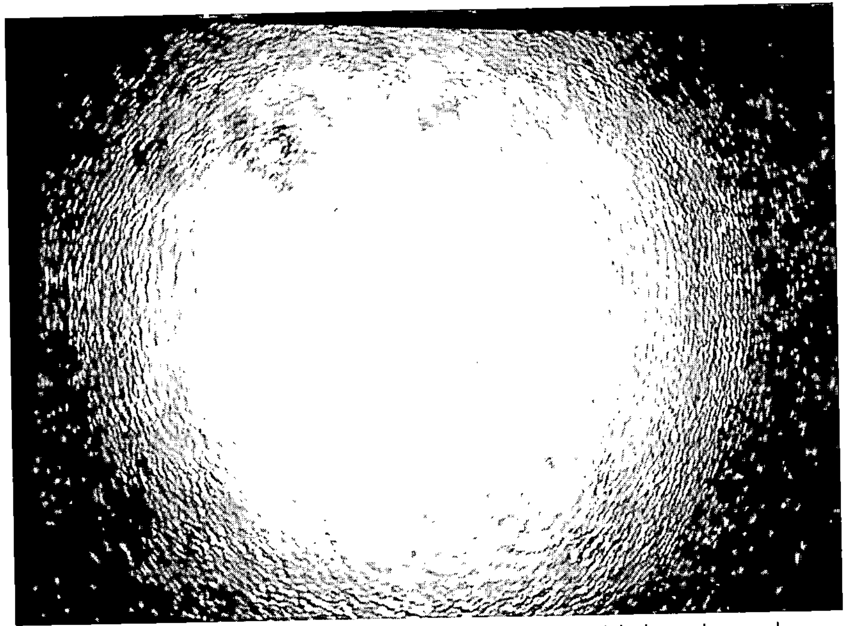

B-16 Beam spread from lens used in spatial filter without pinhole to clean up beam.

First, find the magnification of the lens. Use the following formula:

$$M = 254/f$$

where f is the focal length in millimeters. So for the 9 mm lens:

$$M = 254/9$$
$$= 28$$

The magnification of the lens is 28×. Use that information and the following formula to determine the proper pinhole diameter.

$$D = 2(.6\lambda/Md)$$

where D = diameter of pinhole in μm
M = magnification of lens
λ = wavelength of laser in nm
d = diameter of unspread laser beam

using a 1 mm diameter beam:

$$D = 2((0.6 \times 632.8 \times 10^{-9})/(28 \times 0.001 \text{ m}))$$
$$= 0.000027 \text{ or } 27\ \mu\text{m}$$

Airy disc

It would be nice if we knew exactly the diameter of the pinhole. Measuring something of the order of 0.001 inch (25 μm) can be pretty daunting without specialized equipment. Fortunately, there is a method using what's called an *Airy disc,* named after Sir George Airy, an astronomer who derived a formula in 1835. He used the formula to calculate the resolving power of telescopes. For the applications in this book, the formula is:

$$\text{Diameter of pinhole} = \frac{2.44 \times (632.8 \times 10^{-9}) \times \text{distance}}{\text{diameter of Airy disc}}$$

All measurements must be in meters. See FIG. B-17. Set up the laser so that the unexpanded beam passes through the pinhole. The distance *d* is 1 foot, which equals 0.304 meters. On the screen, observe a diffraction pattern. It is an Airy diffraction pattern, which is a central disk of light with rings around it (as in FIG. B-13). As an example, assume you have the setup in the figure, and the diameter of the disk is 7⁄16 of an inch. Convert that to meters, which is 0.011 meter. Plugging this into the equation gives:

$$\text{Diameter of pinhole} = \frac{2.44 \times (632.8 \times 10^{-9}) \times 0.304}{0.011}$$

$$= 0.000042 \text{ or } 42\ \mu\text{m}$$

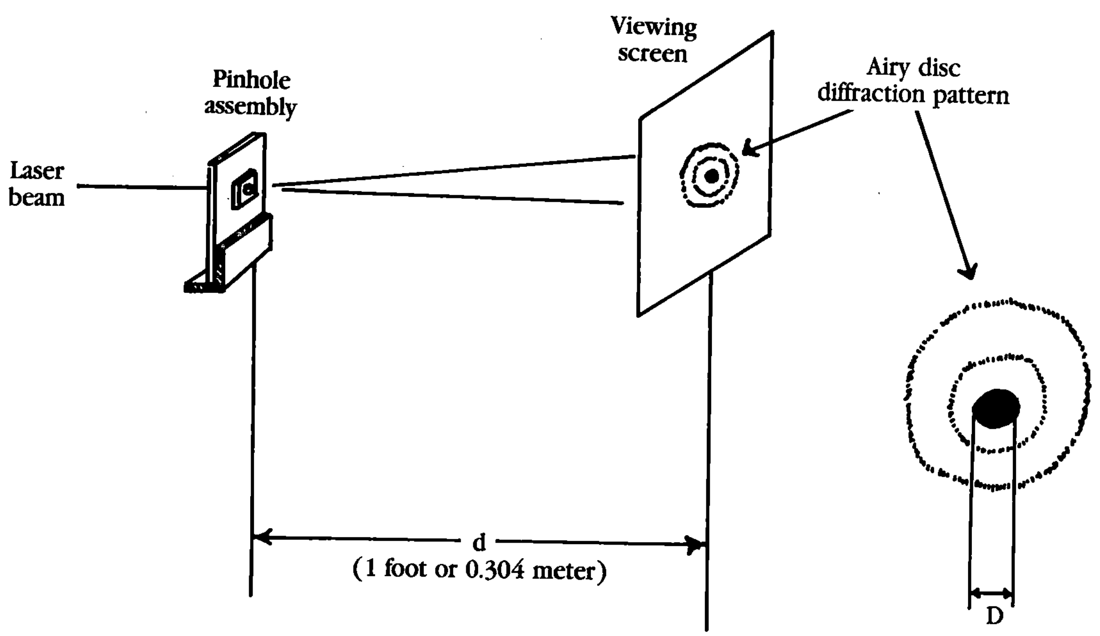

B-17 Airy disc pattern created from pinhole, which is used to calculate diametor of pinhole.

CUTTING GLASS

To prevent glass from breaking off of the cut line, making the piece unusable, a local glazer advised me to spray the wheel of the glass cutter with a lightweight oil, like WD-40 (see FIG. B-18), before making the cut. Buy a new glass cutter similar to the one in the figure and a can of WD-40. Before making the cut, spray the wheel on the cutter with a little WD-40. Using firm pressure, press down on the glass with the cutter and trace the line you want to cut. Only make one pass on the glass. To break it, turn the glass over and tap it across the line, or put pressure on the cut by lifting one end of the glass and applying

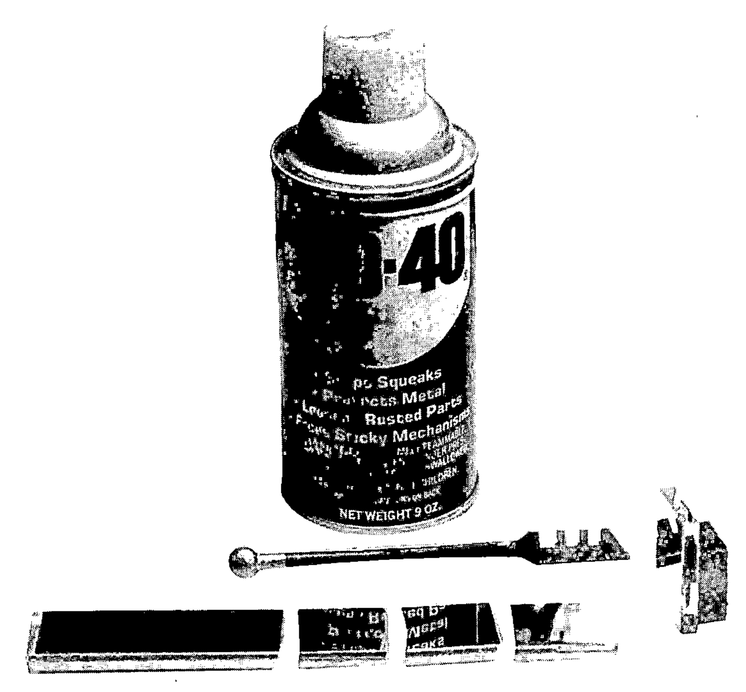

B-18 Photo of glass cutter, WD-40 oil, mirrors and transfer mirror.

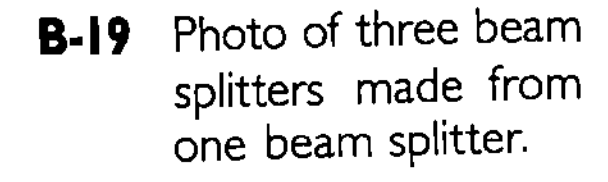

B-19 Photo of three beam splitters made from one beam splitter.

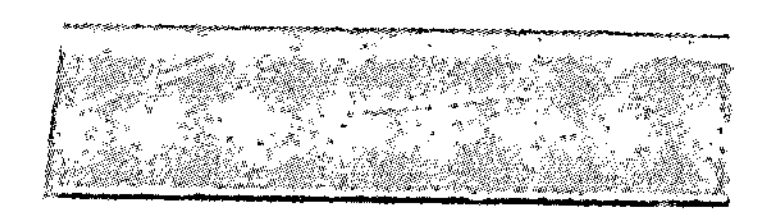

pressure on the line. In either case, always turn the glass over so the glass breaks away from you.

Using this technique, I was able to make what I considered to be some difficult cuts. The mirrors pictured in FIG. B-18 are ¼ inch thick and 1 inch wide. I was able to cut 1-inch-long mirrors off this surplus piece to make a number of transfer mirrors. Figure B-19 shows three beam splitters I made from one purchased beam splitter from Edmund.

Sources

Holography is still an emerging technology, undergoing improvements in methods, materials, and equipment. It's very hard to produce a book that is current or remains current for any length of time. Inasmuch as this is the case, this book remains focused on the basic procedures and methods. I cover a few emerging technologies simply because they are available and are just as easy or easier to use and understand as their predecessors.

PUBLICATIONS

To remain current in the holography field, keep in touch with the magazines, institutions, industries, and publications catering to holography. Following is a list of newsletters, magazines, and other publications you might consider getting a subscription for or becoming aquainted with.

Holographics International
BCM-HOLOGRAPHICS
London WC1N 3XX
England
(081) 642-8381
U.K.-based magazine. Published four times yearly. Excellent source of current events in holography.

The Hologram
P.O. Box 9035
Allentown, PA 18105
(215) 434-8234
Free subscription. Newsletter on holography that contains tips and newsworthy items.

Holography Workshops
Lake Forest College
Lake Forest, IL 60045
(708) 234-3100
Widely reknowned for publication of International Symposium on Display Holography. The books are a compilation of articles by numerous authors and holographers on holography. Well worth cost to purchase as tips, tests, and

methodology, all pertaining to holography, are published. Excellent resource material for serious students of holography. In addition, summer workshops in holography are given under the direction of T. H. Jeong.

Holosphere
MUSEUM OF HOLOGRAPHY
11 Mercer St.
New York, NY 10013
(212) 925-0581
Subscription is $30 per year. Magazine published quarterly by the Museum of Holography. Contains articles, tips, and newsworthy items related to holography.

Laser News
P.O. Box 42083
San Francisco, CA 94101
Subscription rate is $15 per year for four issues. Newsletter/magazine on holography. Often has foil-type hologram enclosed or on cover. Contains tips and newsworthy items related to holograhy.

Leonardo
P.O. Box 75
1442A Walnut St.
Berkeley, CA 94709
(412) 845-8298
Leonardo is published by the Journal of International Society for the Arts, Science, and Technology. Recently, the magazine published one issue exclusively on holography. In this issue, many well-known holographers have written incisive articles on holography. Reportedly, the magazine plans to carry more information and might publish another issue exclusively on holography.

Photonics Spectra
LAURIN PUBLISHING, INC.
Bershire Common
P.O. Box 1146
Pittsfield, MA 01202-9933
Free subscription to qualified readers.

SPIE (Society of Photo-optical Instrumentation Engineers)
P.O. Box 10
Bellingham, WA 98227-0010
Publishes monthly journal *Optical Engineering*. Often contains pertinent information of lasers and optics. Also publishes Proceedings of SPIE *Practical Holography* series. The book is a compilation of articles by numerous authors and holographers on holography. Excellent resource material for serious holography students.

VIEWING HOLOGRAMS

Here is a partial listing of places in the United States you can visit to look at first-class holograms.

THE MUSEUM OF HOLOGRAPHY
11 Mercer St.
New York, NY 10013
(212) 925-0526

MUSEUM OF HOLOGRAPHY/CHICAGO
1134 W. Washington Blvd.
Chicago, IL 60606
(312) 226-1007

THE HOLOS GALLERY
1792 Haight St.
San Francisco, CA 94117
(415) 861-0234

THE WHOLE PICTURE
634 Parkway
Gatlinburg, TN 37738
(615) 436-3650

HOLOGRAPHIC VISIONS
300 S. Grand Ave.
Los Angeles, CA 90071
(213) 687-7171

ELUSIVE IMAGE
West End Marketplace
603 Munger, Ste. 316
Dallas, TX 75202
(214) 720-6060

ELUSIVE IMAGE
Travis Walk
4514 Travis St., Ste. 114
Dallas, TX 75202
(214) 720-6060

LIGHT WAVE
455 E. Illinois St.
Chicago, IL 60611
(312) 204-5344

LIGHT WAVE
Woodfield Mall D-208
Schaumburg, IL 60173
(312) 204-5344

NEWPORT HOLOGRAM
Lido Marina Village
3412 Villa Oporto, Ste. 2
Newport Beach, CA 92663
(714) 675-1337

SURPLUS LASER AND COMPONENT SUPPLIERS

MEREDITH INSTRUMENTS
6403 N. 59th Ave.
P.O. Box 1724
Glendale, AZ 85301
(602) 934-9387

ALLEGRO ELECTRONIC SYSTEMS
3 Mine Mountain Rd.
Cornwell Bridge, CT 06754
(203) 672-0123
Laser surplus house that carries used laser tubes, power supplies, flashtubes, front-surface mirrors, optical flats, and quarter-wave plates. Free catalog.

LASER RESALE, INC.
54H Balcom Rd.
Sudbury, MA 01776
(508) 443-8484
Laser surplus house that carries a variety of lasers including high-power HeNe. Occasionally has surplus ruby rods.

MWK INDUSTRIES
1440-3B S. State College Blvd.
Anaheim, CA 92806
(800) 356-7714 or (714) 956-8497

DICK ANDERSON
2883 West Royalton Rd.
Broadview Heights, OH 44147
(216) 237-0656
Red ruby laser rods, flashtubes, and power supplies.

TIMELINE, INC.
1490 W. Artesia Blvd.
Gardena, CA 90247
(800) 872-8878 or (213) 217-8712
Predominately an electronic surplus house, but occasionally carries reasonably priced lasers. Recently listed 30 mW argon laser for $499, less the power supply, which is a very good price considering argon lasers typically cost a few thousand dollars.

GOTT ELECTRONICS
2227 DuFlour Ave.
Redondo Beach, CA 90278
(213) 370-6287
Predominately an electronic surplus house, but occasionally carries reasonably priced lasers. Recently listed new NEC 10 mW HeNe laser with power supply for $200.

H&R CORPORATION
401 E. Erie Ave.
Philadelphia, PA 19134-1187
Predominately an electronic surplus house, but occasionally carries reasonably priced lasers. Recently listed 2 mW HeNe laser for $105. Power supply is $95. Carries some optical components.

OPTICAL COMPONENTS

EDMUND SCIENTIFIC
101 E. Gloucester Pike
Barrington, NJ 08007-1380
(609) 573-6250
Carries large selection of optics, lenses, beam splitters, front-surface mirrors, inexpensive gratings, and polaroid material. Sells a variety of lasers from different laser manufacturers.

ROLYN OPTICS COMPANY
706 Arrow Grand Circle
Covina, CA 91722-2199
Carries large selection of optics.

LASER MANUFACTURERS

MELLES-GRIOT
1770 Kettering St.
Irvine, CA 92714
(714) 261-5600
Manufactures and sells large variety of optical components, laser tubes, lasers, power supplies, and laser diodes.

AEROTECH
101 Zeta Dr.
Pittsburgh, PA 15238
(412) 963-7459
Manufactures and sells large variety of optical components, laser tubes, lasers, and power supplies.

NEWPORT CORPORATION
P.O. Box 8020
18235 Mt. Baldy Circle
Fountain Valley, CA 92728-8020
(714) 963-9811
Manufactures and sells large variety of optical components, laser tubes, lasers, and power supplies.

SIEMENS CORPORATION
186 Wood Ave. South
Iselin, NJ 08830
(201) 906-4382
Manufactures laser tubes, lasers, and power supplies.

POWER TECHNOLOGY, INC.
P.O. Box 9769
Little Rock, AR 72219
(501) 568-1995
Manufactures and sells power supplies.

TOSHIBA AMERICA, INC.
1220 Midas Way
Sunnyvale, CA 94086
(408) 739-0560
Manufactures and sells visible laser diodes.

HOLOGRAPHIC FILM

INTEGRAF
745 N. Waukegan Rd.
Lake Forest, IL 60045
(312) 234-3756

JOSEPH MERRITT CO.
650 Franklin Ave.
Hartford, CT 06114
(203) 278-2400

CRIMSON CAMERA
325 Vassar St.
Cambridge, MA 02139
(617) 868-5150

CLAUS GELOTTE, INC.
411 Waverly Oaks Rd.
Waltham, MA 02154
(617) 899-2366

KEYSTONE SCIENTIFIC CO.
P.O. Box 22
Thorndale, PA 19372
(215) 384-8092

PHOTOCOPY WAREHOUSE
468 Armour Circle, N.E.
Atlanta, GA 30324
(404) 875-9956

COOK & WELL
50 Route 9
Morganville, NJ 07751
(201) 536-3964
Sells Agfa-Gevaert holographic film and plates.

IMAGES COMPANY
P.O. Box 313
Jamaica, NY 11419
(718) 706-5003

IMAGES CO. SOUTH
P.O. Box 2251
Inverness, FL 32650
(904) 344-8540

JORDON LASER
62 Enterprise Drive
Ann Arbor, MI 48103
(313) 761-4044

LASER LABS
8000 W. 110, Suite 115
Overland Park, KS 66210
(912) 451-9270

ALPHA PHOTO PRODUCTS
P.O. Box 23955
985 Third St.
Oakland, CA 94623
(415) 893-1436

X-RAY SUPPLY CO.
1121 South Central Ave.
Glendale, CA 91204
(213) 247-5273

NEWPORT RESEARCH CORP.
18235 Mt. Baldy Circle
Fountain Valley, CA 92708
(714) 962-7701

MISCELLANEOUS

ARBOR SCIENTIFIC
P.O. Box 2750
Ann Arbor, MI 48106-2750
(800) 367-6695
Sells instructional scientific kits. Carries small amounts of holographic supplies: film, chemistry, sorbathane balls, holographic safelights, and some optics and lasers.

IMAGES COMPANY
P.O. Box 313
Jamaica NY 11419
(718) 706-5003

IMAGES CO. SOUTH
P.O. Box 2251
Inverness, FL 32650
(904) 344-8540

Sells electronic and optical components, sorbathane pads, steel stock, magnets, green transparent plastic for safelights.

PHOTOGRAPHER'S FORMULARY
P.O. Box 5105
Missoula, MT 59806
(800) 777-7158 or (406) 543-4534
Sells holographic development kits. Carries variety of chemicals and compounds for photography and holography.

Glossary

acetate A nonflammable material used as a base for film.

achromatic Black and white.

additive color mixing Two or more wavelengths superimposed upon one another to create additional colors.

agitation Method to keep fresh processing solution in contact with the emulsion surface during processing.

Airy disc The central maximum of a diffraction pattern formed when a parallel monochromatic light passes through a pinhole.

antihalation backing (AH) A dark, light-absorbing material placed on or within the base of a plate or acetate film between the base and emulsion to absorb any light that passes through the emulsion from reflection back into the emulsion.

angstrom A unit of length equal to one millionth of a millimeter (0.1 nanometer or 10^{-10} meter). Its most common use is to measure wavelength in the visible-light spectrum. Abbreviated A.

ASA Speed rating denoting the photographic emulsion's sensitivity to light. Devised by the American Standards Association. The higher the ASA number, the faster the film.

attenuation Reduction of light intensity, or loss. It can be measured in optical density or decibels as well as percentage or fraction of light loss.

attenuator An optical element that transmits only a given fraction of incident light.

average power The average level of power in a series of pulses. It equals pulse energy times the number of pulses divided by the time interval.

ballast resistance The series resistance required to limit current flow and obtain a stable electric discharge.

beam splitter An optical device that divides incident light into two separate beams, one reflected and one transmitted.

bleach Used in holography film processing to change an absorption hologram into a phase hologram to improve efficiency and brightness.

Bragg diffraction Diffraction that is reinforced by reflection of a series of regularly spaced planes, which correspond to a particular wavelength and angular orientation. The angle at which this reinforcement occurs is Bragg's angle.

Brewster's angle The polarizing angle of incidence at which the reflected and refracted rays of light are perpendicular to each other.

chromatic aberration Lens or hologram irregularity due to the shifting position of each frequency. If severe enough, the resulting hologram will appear blurred because of the lack of color registration.

circular polarization Light polarized so the polarization vector rotates periodically without changing absolute magnitude, describing a circle. Circularly polarized light can be considered as two, equal-intensity linearly polarized beams, one 90 degrees in phase ahead of the other.

cladding A sheathing or cover of lower refractive index material intimately in contact with the core of higher refractive index material. It provides optical insulation and protection to the total reflection interface.

coating Material applied to one or more layers of the surface of an optical element to change the way it reflects or transmits light.

coherence Alignment of phase and wavelength of light waves with respect to one another to a point where waveforms produce an interference pattern when combined, as in an interferometer.

coherence length The shortest difference between wavefronts where interference vanishes. This is analogous to the minimum optical path difference in a Michelson interferometer where the interference fringe pattern disappears.

collimate The act of making light rays parallel.

color temperature Way of expressing the color quality of a white light source. The color temperature scale is usually measured in degrees Kelvin.

concave Curving inward, so the central part is deeper than the outside.

continuous wave Emitting a steady beam.

convex Curvature that resembles the outside of a ball.

core The high refractive index material in the center of an optical fiber.

critical angle The maximum angle at which light can be propagated within an optical fiber.

cycles per second Abbreviated cps, it is the obsolete expression for the number of oscillations a wave makes in one second. It is another word for *frequency* (abbreviated Hz).

decibel A logarithmic comparison of power levels. It is abbreviated dB and defined as the value: $10 \log_{10} (P_2/P_1)$, i.e., 10 times the base 10 logarithm of the ratio of the two power levels.

density General descriptive term for the amount of photosensitive material deposited in an emulsion due to exposure and development. Measurement is a logarithm of the opacity (light-stopping ability) of the emulsion.

diffraction The scattering of light or spreading of light waves when they pass an edge or slit.

diode An electronic device that preferentially conducts current in one direction but not in the other. Semiconductor diodes contain a p-n junction between regions of different doping that allows current to flow in one direction, but not the other. Diodes can be made to emit light, as in laser diodes.

divergence The angular spreading of a beam of light with distance ($D = 2L \sin \theta$).

Doppler broadening The broadening of the spectral line width of the laser caused by the random thermal motion of the emitting gas atoms.

electroluminescence Emission of light energy by application of electrical energy to light-sensitive material.

electro-optic The interaction of light and electric fields, typically changing the light wave. Used in some modulators, Q-switches, and beam detectors.

electromagnetic radiation Waves made up of oscillating electric and magnetic fields, perpendicular to one another and propagating at the speed of light. The waves can also be viewed as photons or quanta of energy.

electronic transition A change in the energy level of an electron.

fiber An optical transmission element consisting of a fiber core and cladding.

frequency For light waves, the number of wave peaks per second that pass a point. Measured in hertz (cycles per second).

FWHM Full width at half maximum; a convenient measure of Doppler-broadened line widths and cavity resonances.

gallium aluminum arsenide A semiconductor compound used in LEDs and diode lasers. Its chemical formula is GaAlAs.

gallium arsenide A semiconductor compound used in LEDs, diode lasers, and electronic components. Its chemical formula is GaAs.

gelatin A natural protein used as the transparent medium to hold light-sensitive material in suspension, which binds them to film or plate but allows the passage of processing solutions.

grain Size and pattern of photosensitive silver, which often clumps together in an emulsion when silver material has been exposed and developed.

hertz The unit of measurement for frequency.

incoherent Light waves that are out of phase, multidirectional and/or multifrequency.

index of refraction The ratio of the speed of light in a vacuum to the speed of light in a material.

infrared Invisible wavelengths from about 700 to 2000 nanometers.

injection laser Another name for the semiconductor laser. The name comes from the fact that the injection of current leads to the production of light.

intensity Power per unit solid angle.

interference The sum of the amplitudes of light waves. In destructive interference, the waves cancel; in constructive interference, they add to make more intense light.

irradiance Power per unit area.

joule Measurement of energy equal to 1 watt/second.

junction laser A semiconductor diode laser.

Kelvin Unit of measurement used to indicate the color temperature of light. The Kelvin scale is numerically equal to the absolute scale by adding 273 to degrees centigrade.

laser Acronym for *l*ight *a*mplification by *s*timulated *e*mission of *r*adiation.

LED Acronym for *l*ight-*e*mitting *d*iode. A semiconductor diode that emits incoherent light by spontaneous emission.

light Form of electromagnetic radiation visible to the human eye. It has a range of wavelengths from 400 to 700 nanometers.

light-emitting diode A semiconductor diode (*see* LED).

longitudinal modes Oscillation resonances of laser wavelengths along the length of its cavity. Equal to twice the length of the cavity divided by the integral number of wavelengths. The wavelengths must be close enough to fall within the Doppler-broadened line width of the atomic transitions of neon. Distinct from transverse modes, which are across the width of the cavity.

longitudinal modes spacing The frequency separation of adjacent longitudinal modes.

maser Microwave analog of a laser. Acronym for *m*icrowave *a*mplification by *s*timulated *e*mission of *r*adiation.

micrometer One millionth of a meter, abbreviated μm.

mode A manner of oscillation in a laser. Modes can be longitudinal or transverse.

monochromatic Containing only a single frequency or wavelength.

multimode Containing multiple modes of light. Typically refers to lasers that operate in two or more transverse modes.

n region Part of a semiconductor doped so that it has an excess of electrons as current carriers.

nanometer A unit of length equal to 10^{-9} meter. Its most common use is to measure visible wavelengths.

nanosecond A billionth (10^{-9}) of a second.

neutral-density filter Grey filter which reduces the amount of light without affecting color.

normal (angle) Perpendicular to a surface.

p region Part of a semiconductor doped with electron acceptors in which holes (vacancies in the valence electron level) are the dominant current carriers.

phase Position of a wave in its oscillation cycle.

photodetector A light detector.

photometer An instrument for measuring the amount of light visible to the human eye.

photons Quanta of electromagnetic radiation. Light can be viewed either as a wave phenomenon or a series of photons.

polarization As a light wave travels, the electrical and magnetic fields oscillate in all directions. Polarizing light can be achieved by passing the light through a polarizing filter, which only allows light to pass through the electrical field position that matches the alignment of the polarizing material crystals. This polarizes or aligns the electric and magnetic fields of electromagnetic light waves.

polarization vector A vector that indicates the direction of the electric field in an electromagnetic wave.

polarizer An optical material or device that transmits light of only one polarization.

population inversion The condition when more atoms are in an upper energy level than in a lower one. Population inversion is required for laser action.

pumping The method used in a laser to generate population inversion.

Q switch A device that changes the Q (quality factor) of a laser cavity to produce a short, powerful pulse.

radiant flux Instantaneous power level in watts.

radiometer An instrument to measure power (watts) in electromagnetic radiation. Distinct from a photometer.

rays Straight lines that represent the path taken by light.

recombination Capture of a conduction-band electron by a "hole" in a semiconductor, causing the conduction band to drop a vacancy in the valence band.

refraction The bending of light as it passes between materials of different refraction indices.

refraction index The ratio of the speed of light in a vacuum to the speed of light in a material.

semiconductor diode laser A laser in which recombination of current carriers at a p-n junction generates stimulated emission.

speckle Coherent noise produced by laser light. It gives a mottled appearance to holograms viewed in laser light.

spontaneous emission Emission of a photon without outside stimulation when an atom or molecule drops from a high-energy state to a lower one.

stimulated emission Emission of a photon that is stimulated by another photon with the same energy. This process makes laser light.

squeegee Implement used to squeeze water off of plate or film emulsion. It can be a roller or a flat, rubber type that functions like a windshield wiper blade.

test strip Emulsion that is given a range of exposures to test for the correct density.

transition Shift between energy levels.

transverse modes Modes across the width of a laser medium. Distinct from longitudinal modes, which are along the length.

tunable Adjustable in wavelength.

visible light Electromagnetic radiation visible to the human eye at wavelengths of 400 nm to 700 nm.

waveguide A structure that guides electromagnetic waves along its length. An optical fiber is an optical waveguide.

wavelength The distance an electromagnetic wave travels during one cycle of oscillation. Wavelengths of light are measured in nanometers.

wetting agent Chemical compound that reduces the surface tension of water. It is added to the final rinse of a hologram to improve water drainage off the plate as it is drying to prevent drying marks. It also reduces drying time.

Index

S

T

U

V

W

Y